12/86

D0856354

THE MANIPULATION OF
AIR-SENSITIVE COMPOUNDS

THE MANIPULATION OF AIR-SENSITIVE COMPOUNDS

SECOND EDITION

D. F. SHRIVER

Northwestern University
Evanston, Illinois

M. A. DREZDZON

Amoco Chemicals Company
Naperville, Illinois

A Wiley-Interscience Publication

JOHN WILEY & SONS

New York / Chichester / Brisbane / Toronto / Singapore

Library of Congress Cataloging in Publication Data:

Shriver, D. F. (Duward F.), 1934—
    The manipulation of air-sensitive compounds.

    "A Wiley-Interscience publication."
    Includes bibliographical references and index.
    1. Chemistry—Manipulation.   2. Vacuum technology.
3. Protective atmospheres.   4. Air-sensitive compounds.
I. Drezdzon, M. A. (Mark A.)   II. Title.

QD61.S57   1986   542   86-11012
ISBN 0-471-86773-X

Printed in the United States of America

10 9 8 7 6 5 4 3 2 1

PREFACE

The continued importance of research and development with air-sensitive compounds, coupled with the general lack of publications or undergraduate instruction on the techniques used in this area, prompted us to revise this monograph. The level of presentation is designed to make the book useful to the chemist who is beginning work in the field. In addition, considerable technical data and information are given that should aid chemists at all levels of proficiency in the design of experiments.

In the belief that the beginner needs a selective approach and the more seasoned worker will exercise his ingenuity, we have refrained from presenting brief references to many similar items of equipment, but have attempted to present rational approaches to the common operations involved in the synthesis, separation, and characterization of air-sensitive materials, and to illustrate these with typical apparatus. Whenever possible, these examples are based on equipment with which we have had experience or on firsthand observation of the equipment used by others. This has resulted in a somewhat personalized account, which we hope will ensure reliability without being objectionable to established workers with different preferences in equipment design. The techniques discussed here were largely developed for use with inorganic and organometallic compounds; however, they are generally useful for problems involving gases and air-sensitive solids or liquids, so they find application in all areas of chemistry, as well as biology and physics.

We had two goals in mind while writing this second edition. The first was to bring the text up to date. Although there have been no radical new developments, there has been a steady improvement in technique. Our second objective was to make the book more accessible to the reader interested in a specific technique. Thus more section headings are used and more detailed examples are given.

In the preparation of the first edition of this book aid was received from many people: A. L. Allred, F. Basolo, R. L. Burwell, Jr., E. W. Schlag, Paul Treichel, R. W. Fowells, and Shirley Shriver. Useful suggestions for the second edition were provided by L. Aspry, R. L. Burwell, Jr., P. Bogdan, N. J. Cooper, D. Kurtz, J. Malm, and S. H. Strauss. For aid in proofreading we appreciate the help of Ann Crespi, Cynthia Schauer, Daniel Shriver and Ralph Spindler.

D. F. SHRIVER
M. A. DREZDZON

Evanston, Illinois
Naperville, Illinois
February, 1986

CONTENTS

PART 2 VACUUM LINE MANIPULATIONS

INTRODUCTION

A. General Methods. The most widely used methods for handling air-sensitive compounds are based on the use of an inert gas atmosphere to exclude air. This approach may be further subdivided into (1) those techniques which involve bench-top operations with special glassware (often called Schlenk techniques, Chapter 1) and (2) glove box techniques in which conventional manipulations are performed in an inert-atmosphere box (Chapter 2). Both of these methods may be used to handle large quantities of solids and liquids; however, the Schlenk type equipment is generally much more efficient and more secure from the atmosphere than the dry box. The principal advantages of the dry box lie in its use for intricate manipulations of solids and for the containment of radioactive and/or highly poisonous substances.

More rigorous exclusion of air and the quantitative manipulation of gases is achieved in a previously evacuated apparatus. This vacuum line method is outlined in Chapter 5, where the basic operations are described, such as quantitative transfer and pressure–volume–temperature measurement of gases, trap-to-trap separations of volatile materials, and the use of vapor pressures to characterize substances. Succeeding chapters (6–8) describe in detail the individual vacuum line components such as stopcocks, joints, diffusion pumps, and manometers. Many accessory items such as spectroscopy cells, vapor-pressure thermometers, and gas-chromatography apparatus are described in Chapter 9. This set of chapters, plus supplementary material in the Appendixes, form a guide to modern glass vacuum line practice. This technique has been used successfully in the synthesis and manipulation of hydrides, halides, and many other volatile substances. However, glass vacuum systems are not appropriate for hydrogen fluoride and some other reactive fluorides. These are best handled in the metal or fluorocarbon apparatus described in Chapter 10. The strengths of these various techniques for handling air-sensitive materials are summarized in Table 1.

Table 1. Comparison of Inert-Atmosphere and Vacuum Line Techniques

Technique	Exclusion of Air	Quantities[a]	Outstanding Features
Schlenk and syringe	Good to very good	Medium to large	Rapid and easy manipulation of solutions; simple transfer of solids
Glove box	Poor to very good	Small to very large	Intricate transfer operations; containment of poisonous and radioactive substances
Vacuum line	Very good to excellent	Very small to medium	Quantitative gas handling; transfer and storage of volatile solvents without contamination

[a]Very small, about 1 mg; small, about 100 mg; medium, several grams; large, hundreds of grams; very large, kilograms.

B. Planning the Experiment.   Whatever general technique is used, it is important to plan each new experiment in detail. Impasses can be avoided by sketching the setup at each step with due attention paid to the method for transferring materials. If measurements are involved, a simple estimate of the potential errors should precede the newly planned experiment to ensure that results will be determined with meaningful accuracy. Frequently, such an estimate will suggest conditions and apparatus designs which will minimize the errors without introducing any new complications.

C. Apparatus Design.   Sometimes it is necessary to design and construct special apparatus. Three criteria for a good design are: (1) Is it the simplest design consistent with its intended purpose? (2) Is it robust? (3) Is it easy to clean? To ensure that an item is sturdy it should be designed so that stresses are not concentrated in small sections of glass. For permanent installations, such as a vacuum line, all heavy items are individually supported, while the lighter sections are clamped at the minimum number of points which suffice to support the apparatus and avoid leverage. Generally, portable glassware should be designed as one strong, compact unit which may be supported near a point of balance. Finally, the materials of construction must withstand the chemicals and solvents which are to be handled. The most frequent problems arise with stopcock greases, waxes, rubbers, and plastics. These problems may be minimized by selecting materials on the basis of the information on chemical and solvent resistance which is presented in Chapter 8 and Appendix III. To aid in the proper choice of metals, information on their corrosion resistance is given in Chapter 10 and Appendix IV.

D. Check-Out and Troubleshooting. Even though an experiment may be well conceived and the equipment properly designed, it is usually necessary to remove small defects in workmanship from a newly constructed or altered apparatus. Therefore, the experimentalist must be prepared to check for faults whenever a change has been made. For example, it is necessary to carefully check a new vacuum apparatus or glove box for leaks.

Inert-Atmosphere Techniques

1

BENCH-TOP
INERT-ATMOSPHERE
TECHNIQUES

Glass apparatus from which the atmosphere is excluded by an inert gas is very widely used in the preparation of air-sensitive inorganic, organic, organometallic, and biochemical materials. The techniques are often adaptations of common laboratory operations, and therefore both cost and complexity are minimized. These methods are well suited for handling solutions, and also may be used for solids and reactive gases. However, glove boxes are usually preferred for intricate solids handling operations, and vacuum lines are superior for quantitative gas handling. Further comparisons of the various inert-atmosphere and vacuum line techniques are given in the Introduction.

Attention also is directed to other chapters which are useful adjuncts to the present one: "Inert Gases and Their Purification" (Chapter 3), "Purification of Solvents and Reagents" (Chapter 4), and "Joints, Stopcocks, and Valves" (Chapter 8). The solvent and chemical resistance of elastomers, which are used in O-ring joints and septa, are discussed in Chapter 8 and Appendix III.

1.1 TECHNIQUES FOR PURGING AND DRYING APPARATUS

A. Initial Purge. There are two common methods for removing atmospheric gases from inert-atmosphere glassware. The first of these methods involves the evacuation of the apparatus followed by filling with an inert gas. Several cycles of pumping followed by filling with inert gas are often performed to avoid the need to employ high-vacuum techniques. If f is the fraction of gas remaining after

7

evacuation of the apparatus, and if there are no leaks in the apparatus, n repetitions of the pump–and–fill process will reduce the fraction of atmospheric gases A_f to

$$A_f = f^n \qquad (1)$$

The second method for removing atmospheric gases involves sweeping air out of the apparatus by a flush of inert gas. The factors which influence this type of process are discussed in detail in Chapter 2. For the present, it is adequate to note that a continuous flush of inert gas, which pushes the atmospheric gases from one extreme of the apparatus to an outlet on another extreme, as in Fig. 1.1, is most efficient. This so-called plug flow is difficult to achieve with single-neck flasks and similar apparatus. A flow of inert gas which bypasses part of the apparatus is relatively inefficient.

B. Purging an Open Vessel.   In many operations described in this chapter, it is necessary to open the apparatus briefly while an inert-atmosphere flush is maintained out of the opening to minimize the entrance of air. Even under conditions in which there is little turbulence in the flowing inert gas stream,

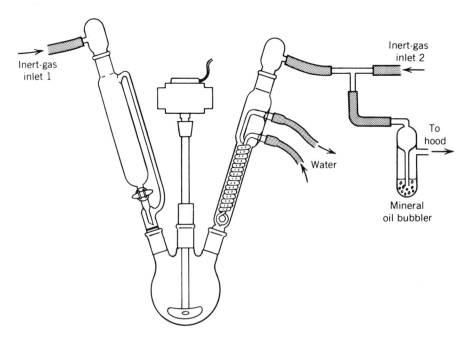

Fig. 1.1.   Three-necked reaction flask with dropping funnel, stirrer, and reflux condenser. With the dropping funnel stopcock in the open position, a flow of inert gas from inlet 1 through the mineral oil bubbler efficiently purges the apparatus of atmospheric gases. During reaction and subsequent cooling of the reaction mixture, a slow flow of inert gas from inlet 2 through the mineral oil bubbler prevents atmospheric gases from backing up into the system, and this configuration minimizes exposure of the reaction mixture to impurities in the inert gas.

there will be countercurrent diffusion of atmospheric gases. An estimate of the flow rate L necessary to maintain the partial pressure P_0 of an atmospheric component at some desired partial pressure P inside the apparatus is given by the equation[1]

$$L = -(AD_{1,2}/X)ln(P/P_0) \tag{2}$$

where A is the cross-sectional area of the tube through which the gas is issuing, X is the length of this tube, and $D_{1,2}$ is the diffusion coefficient of the impurity gas 1 in the inert gas 2. In round numbers, the diffusion coefficient of oxygen in nitrogen is $0.2 \text{ cm}^2/\text{s}$ at 1 atm total pressure, and a desirable partial pressure for oxygen within the apparatus might be 10^{-3} torr. If we assume that the neck of a flask is 1.5 cm in diameter and 5 cm long, Eq. (2) indicates that a nitrogen flow rate of only $50 \text{ cm}^3/\text{min}$ would be required to maintain this low level of oxygen. Experience teaches us that far larger flow rates are required; something on the order of several liters per minute would be typical under these conditions. Even though Eq. (2) does not appear to apply to the rather large openings common to many preparative-scale apparatus designs, it is probably much more accurate for the less turbulent flow of gas from smaller-diameter tubing. The general form of the equation is sensible, suggesting that the entrance of air is reduced by a long-necked flask with a small cross section for this neck.

C. Manifold for Inert Gas and Vacuum.

Pure inert gas, generally nitrogen but sometimes argon or helium, is required for the operations described in this chapter. As described earlier, equipment is filled with the gas by purging with a large volume of gas, or evacuation followed by filling with the gas. To accomplish these purging operations in an efficient manner it is handy to have an inert-gas manifold with several heavy-walled flexible vinyl or rubber tubes attached to provide inert gas to separate pieces of apparatus. A mineral oil bubbler, or occasionally a mercury bubbler, is attached to the gas outlet on the apparatus to protect against excessive pressure.

When the pump–and–fill technique is used, a more complex manifold for the distribution of inert gas and vacuum is generally employed, in conjunction with a liquid nitrogen-cooled trap, mechanical vacuum pump, and pressure release bubblers (Fig. 1.2).

This manifold can be used to purge several pieces of apparatus at once, and the two-way stopcocks or valves provide a ready means of switching between inert gas and vacuum. Sources of purified inert gas and vacuum are attached to this manifold. An oil bubbler on the inert gas inlet serves as an approximate flow indicator. The inert gas is controlled at the tank with a high-quality dual-stage diaphragm regulator which is designed for good regulation around 3 psig (915 torr). A pressure release bubbler is often included on the inert gas line. Glass stopcocks attached to these manifolds will become dislodged by the small positive pressure of inert gas. Therefore, it is essential to secure these stopcocks, and

[1]G. Antos, Ph.D. Thesis, Northwestern University, 1973.

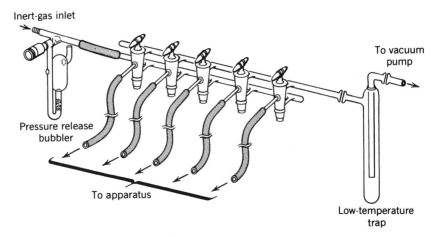

Fig. 1.2. Manifold for medium vacuum and inert gas. A low-temperature trap, on the right side of the figure, is used in the vacuum line to protect the pump from harmful vapors. When the apparatus is being filled with gas or purged with inert gas, the valve on the pressure release bubbler (which contains a check valve to prevent oil from backing up into the line) is opened to avoid excess pressure which would blow the apparatus apart. Often a mineral oil bubbler (not shown here) is connected in line with the inert-gas source to provide visual indication of the inert-gas flow.

all others used on inert-atmosphere equipment, with high-quality plug retainers.[2]

Details on the purification of inert gases can be found in Chapter 3. The source of vacuum is generally a rotary mechanical vacuum pump. The pump is protected from chemicals and solvent vapor by means of a trap cooled by Dry Ice or, preferably, liquid nitrogen. **CAUTION: A liquid nitrogen trap must never be connected to a manifold where the vacuum source has been turned off. Failure to remove a liquid nitrogen trap from a manifold after shutting off the vacuum will result in the condensation of liquid air in the trap. If warmed up, this liquid air will evaporate and may pressurize the apparatus, presenting an extreme explosion hazard.** It is desirable for this trap to be deep enough to extend to the bottom of a 1-L Dewar, thus permitting long-term cooling. The trap should be easily removed so that accumulated condensables can be readily (and frequently) discarded.

D. Purging Syringes. Syringes are conveniently purged from an inert-gas source such as a tube with a rapidly flowing inert-gas stream or a special septum attached to the inert-gas manifold. Two or three cycles of filling, removing the

---

[2]Excellent retainers (K-809000), which must be used with stopcock plugs from the same manufacturer, are available from Kontes Glass Co. Vineland NJ.

needle from the inert-gas source, and expelling the gas are sufficient. If the syringe is going to be used to remove liquid from a flask which is being rapidly flushed, the inert gas may be drawn from the flask and expelled outside of the flask to purge the syringe. When highly moisture-sensitive materials, such as aluminum alkyls and active borohydrides, are being handled, the syringe parts should be stored in a drying oven around 130°C and removed just prior to purging.

E. Drying Glassware. The purging procedures described thus far remove relatively little of the moisture which is adsorbed on glass surfaces. When handling compounds which are highly moisture sensitive, such as aluminum hydrides, borane adducts, and lithium alkyls, it is often important to employ special drying procedures. A drying oven set at around 130°C can be used to bake out glassware for several hours prior to use. This glassware should be flushed with dry inert gas as it cools. To prevent apparatus from sticking together, standard taper joints, stopcocks, and syringes should be completely disassembled before they are placed in the oven.

Alternatively, the glassware may be flushed while it is being heated with a Bunsen burner or heat gun. **CAUTION: When using a Bunsen burner or heat gun to dry glassware, all flammable materials must be removed from the area. If the glassware has been rinsed with a volatile solvent, such as acetone or methanol, to aid the drying process, the residual solvent vapors should be removed from the glassware prior to drying by purging several minutes with inert gas.** The heating is started near the inert-gas inlet and carried along the apparatus toward the gas outlet.

Silylating agents have also been used to treat glassware before handling moisture-sensitive compounds. For example, a chlorine-terminated polydimethyl siloxane is available commercially.[3] These reagents suppress the basicity of the glass and provide a hydrophobic surface.

1.2 ADAPTATIONS OF STANDARD GLASSWARE

Many operations with air-sensitive liquids can be performed in standard taper glassware fitted with a bubbler and inert-gas source. **NOTE: To avoid blowing apparatus apart, stills and reaction pots should be vented to a bubbler before they are heated.** Figure 1.1 illustrates a typical setup for a three-neck reaction flask equipped with a dropping funnel, stirrer, and reflux condenser. This arrangement provides several important features for safe and efficient operation. Inert-gas inlet 1 and the oulet through the mineral oil bubbler are positioned for efficient initial purge. Pressure buildup, especially during heating, is avoided by allowing pressure release through the bubbler. A constant slow flow of inert gas

---

[3]This material is sold under the name Glassclad by Petrarch Systems, Inc., Bartram Road, Bristol, PA 19007.

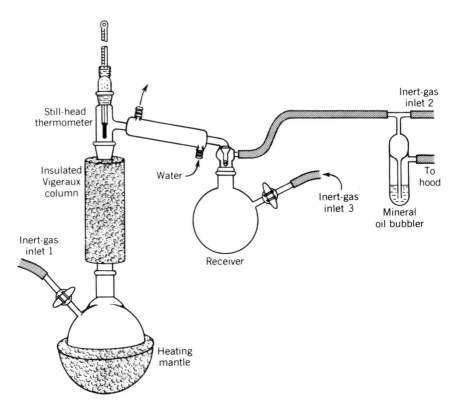

Fig. 1.3. Distillation under inert atmosphere using modified standard taper ware. Note the side-arm added to the standard single-necked still pot and receiver. The sidearm on the still pot allows efficient initial purging of the apparatus, while the sidearm on the receiver allows one to maintain a brisk inert-gas flow over the distilled solvent when the receiver is removed from the still. The apparatus is initially purged as follows: (1) The entire apparatus, except for the receiver, is assembled. (2) Inert-gas inlet 1 is opened which purges the still pot, column, and condenser. (3) The receiver is attached to inert-gas inlet 3 and purged separately. (4) After both the receiver and the main part of the apparatus have been sufficiently purged (about 3–5 min), the receiver is attached to the main apparatus while maintaining the inert-gas flow from both inlets 1 and 3. When the receiver is attached, there will be a vigorous inert-gas flow through the mineral oil bubbler. (5) A slow flow of inert gas is started from inlet 2. The inert-gas flow from inlets 1 and 3 is now terminated. Distillation may now begin. The slow flow of inert-gas from inlet 2 through the mineral oil bubbler will prevent air from backing up into the apparatus while minimizing exposure of the solvent to any impurities present in the inert gas. If the inert-gas flow is maintained through the still pot during the distillation, the efficiency of the separation would be degraded. When the distillation is complete, the inert-gas flow from inlet 3 is resumed before disconnecting the receiver from the apparatus.

through inlet 2, with no flow from inlet 1, prevents air contamination during the course of the reaction while not exposing the entire system to any impurities that may be present in the inert-gas stream. This slow inert-gas flow from inlet 2 also prevents air from being sucked in while the apparatus is cooling after the reaction has been completed. Finally, this arrangement prevents the depletion of highly volatile solvents.

An apparatus for distillation under inert atmosphere, illustrated in Fig. 1.3, incorporates the same features seen in the inert-atmosphere reaction apparatus (Fig. 1.1). The distillation apparatus also contains some slightly modified round-bottom flasks for more efficient initial purging. These modifications will be discussed in greater detail in the following section.

To minimize leakage, joints are either lightly greased or equipped with Teflon sleeves, and are held together with springs, special clips, or rubber bands. When it can be used, grease provides the tightest seal; however, Teflon sleeves are generally used for connections to still pots and similar harsh environments. The various types of stopcock grease, including the new solvent-resistant Teflon-based lubricants, are summarized in Chapter 8. Greased standard taper joints are readily available and are adequate for most purposes; however, O-ring joints are superior for most inert-atmosphere apparatus. The advantages of O-joints are that they are resistant to solvents, and the clamps used with these joints hold the apparatus together much more positively than do the springs and other devices used on standard taper joints. Very satisfactory operation is possible with either standard O-joints or the Teflon-supported O-ring joints (Solv-Seals). These types of joints are described further in Chapter 8. If standard O-joints are used, the O-ring material should match the solvents being handled; again, Chapter 8 should be consulted for details.

The principal drawbacks of the standard apparatus is that it is difficult to flush efficiently and liquid transfer operations are awkward. With little additional complexity, the apparatus can be adapted for efficient purging and solution transfer by syringe or cannula.

1.3 SYRINGE AND CANNULA TECHNIQUES

A. Typical Apparatus.   A simple modification which improves the utility of a one-neck still pot or solvent receiver is a sidearm, which can be used for flushing the flask. This simple modification facilitates the initial flush of the apparatus shown in Fig. 1.3. Furthermore, the sidearm provides several alternatives for the removal of solvents from a receiver, such as flushing the flask while solvent is removed through the standard taper joint, or removal of solvent through the sidearm, as illustrated in Fig. 1.4. The sidearm also permits the flask to be maintained under a constant flush of inert gas when it is attached to another piece of apparatus (Fig. 1.5).

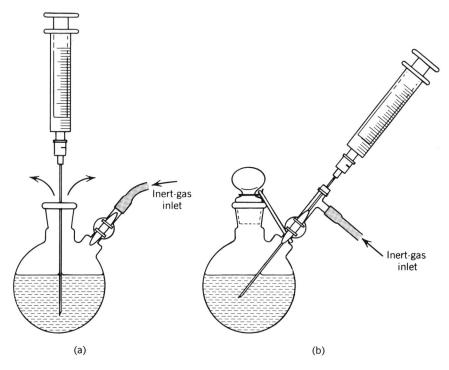

(a)   (b)

Fig. 1.4. Transfer of solvent by syringe from a storage flask with the exclusion of air. (a) Solvent is removed through the neck of the storage flask while maintaining a brisk flow of inert gas from the sidearm. (b) Solvent also may be removed through the sidearm while maintaining 1 atm of pressure in the flask by admitting inert gas through the inlet. In both (a) and (b), the syringe is initially purged by sampling inert gas from the storage flask and expelling this gas outside the storage flask.

B. Septa and Other Closures.   Rubber septa may be attached in a variety of ways, as illustrated in Fig. 1.6. The flat variety is held in place by screw caps, crimped caps, or beveled holders. Various manufacturers provide apparatus with these types of septum closures.[4] Of greater versatility are the sleeve septum stoppers (Fig. 1.7), which can be attached to straight tubes or standard taper joints without any special fixtures.[5]

Two problems with all septa are their sensitivity to solvents and chemicals and leakage through the punctured septum. Flat composite septa minimize these problems. These consist of a central core of a compliant rubber, such as a soft

---

[4]Special septum closures and apparatus with these closures are manufactured by Ace Glass Co., Aldrich Chemical Co., Kontes Glass Co., and Wheaton Glass Co. in the United States, and Sovirel in France.

[5]Sleeve-type septum closures are widely available from chemical and hospital supply houses. An excellent sleeve septum stopper, produced by the Suba Seal Co. in the United Kingdom, is available there from Gallenkamps and from Strem Chemical Co. in the United States.

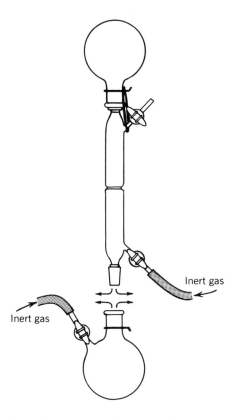

Fig. 1.5. Joining two pieces of equipment under inert-gas flush. Both pieces are initially purged with inert gas separately before joining. Note the use of wire hooks and rubber bands to secure individual pieces of glassware. If the glassware is constructed using O-ring joints, the O-ring joint clamps hold the apparatus together.

silicone rubber, sandwiched between thin sheets of Teflon. The silicone rubber imparts good sealing action, and the Teflon is highly resistant to solvents and to permeation by atmospheric gases. Direct contact of solvents with septa should be minimized and septa should be replaced often. When organic solvents are being used, one must be wary of impurities extracted from septa, joints, and tubing. These extracted impurities may foul reactions and cause considerable confusion in the interpretation of IR and NMR spectra. The prime offenders are hydrocarbon and silicone stopcock greases, dibutyl phthalate and similar plasticizers from Tygon tubing, and various extracts from rubber goods. Some spectral methods for identification of these nuisances are collected in Table 1.1. The pickup of impurities from septa can be greatly minimized by prior extraction with an appropriate hot solvent followed by pumping off the absorbed solvent before use. Septa make poor closures for containers used for the long-term storage of highly air-sensitive materials, since atmospheric gases diffuse through

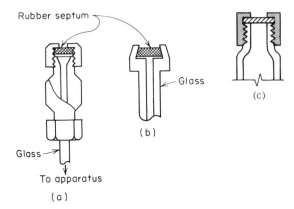

Fig. 1.6. Methods for attaching rubber septa to glass apparatus. (*a*) A modified Swagelok-type fitting. The lip on one end of the fitting is turned down on a lathe, and the septum replaces the ferrules. On the other end, a Teflon front ferrule is used to make the connection with a glass tube. (*b*) An all-glass septum holder. (*c*) Cross section of a plastic threaded cap septum holder on a threaded glass sidearm (manufactured by Wheaton Glass Co.).

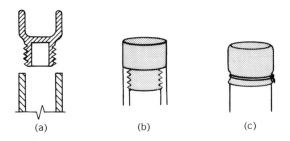

Fig. 1.7. Securing a Suba-seal into a glass tube. (*a*) Cross section of Suba-seal as it is inserted into a glass tube. (*b*) After inserting, the flap on the Suba-seal is folded over the tube. (*c*) The flap may be secured to the tube by wire.

Table 1.1   Spectral Signatures of Impurities from Stopcock Grease and Rubber and Plastic Items

Source	Nature of Impurity	Spectroscopic Identification
Rubber septa or tubing	Dyes, antioxidants, etc.	Vis-UV absorption
Tygon and other flexible poly(vinyl chloride) tubing	Dibutyl phthalate and other plasticizers	NMR: Mult. 7.5, dub. 4.5 Mult. 1.1–1.9, Mult. 0.92 ppm
Stopcocks and joints	Silicone grease Apiezon L	NMR: ca. 0.1 ppm NMR: ca. 1.25, 0.9 ppm

septa and the rubber is likely to undergo chemical breakdown. One possible exception to this statement is the use of unpunctured Teflon-coated septa as seals; but even here one is better off using sealed glass ampules for storage.

C. Syringes.
The principal variations in available syringes are in the design of the tips and plungers. For most preparative-scale work, syringes with removable needles are employed. The removable needle with a female joint hub is attached to a male tapered joint on the syringe. This male Luer tip may be a straight inner joint fashioned from glass or metal or a metal joint with surrounding locking device which holds the needle securely in place. The metal locking variety is preferable because it prevents the detachment of the needle during critical transfer operations and the metal tip is more robust than a glass tip. Glass tips are desirable when highly corrosive liquids are handled. The Luer taper joints are widely adopted but not universal. Nonstandard tapers do not give an air tight seal if mixed with a Luer component. Microliter syringes, which are useful for sampling small volumes of liquids, often have needles which are permanently attached to the barrels.

The standard small syringes are often constructed with individually matched plungers and barrels which are given matching code numbers. These plungers and barrels are not interchangeable with those from other syringes. Larger syringes often have interchangeable parts. Leakage of air past the plunger is a constant problem with these types of syringes. When a light coating of mineral oil on the plunger can be tolerated, this provides an effective means of reducing the leakage. It also is possible to reduce the entrance of air past the plunger by forcing the solution into the syringe with a small positive pressure rather than sucking the material into the syringe by pulling on the plunger. For example, the three-needle technique illustrated in Fig. 1.8 may be used to fill a syringe under controlled positive pressure. The various methods of forcing the solution into the syringe are done with the greatest control by using a metal syringe holder which limits plunger travel and thus avoids the possibility of the plunger popping out of the syringe (one such holder is illustrated in Fig. 1.9). It is especially helpful to use a metal syringe holder when pyrophoric materials, such as neat aluminum alkyls, are transferred by syringe. The metal holder also has the convenience of permitting one-handed operation of the syringe.

Tighter seals between the plunger and barrel are achieved with so-called gas tight syringes in which a Teflon-tipped plunger or an O-ring-equipped plunger is employed. Very good leak resistance is also displayed by the inexpensive "disposable syringes," which consist of a polypropylene shell and a rubber-tipped plunger. Unfortunately, the rubber swells and sticks in the barrel when most organic solvents are transferred; but these disposable syringes are excellent for work with aqueous solutions.

D. Syringe Needles.
Syringe cannulae, which will be referred to as needles throughout this discussion, are most commonly constructed from stainless steel tubing fitted to a metal or plastic hub. Chromium-plated brass hubs are perhaps

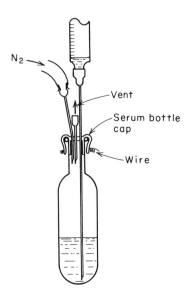

Fig. 1.8. The three-needle technique for withdrawing liquid from a storage tube. The syringe is initially purged by sampling inert gas from the tube and expelling the gas outside the tube. To fill the syringe, the vent is briefly covered with a finger, forcing the solution into the syringe.

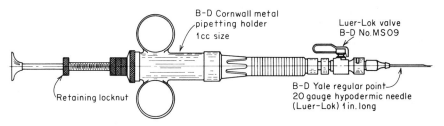

Assembled hypodermic syringe in metal pipetting holder

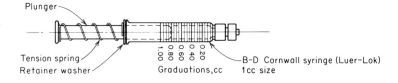

Plunger assembly

Fig. 1.9. Assembled hypodermic syringe in metal pipette holder.

the most common variety. Leaks in the Luer taper joint between the needle and syringe body may be minimized by the use of a small amount of stopcock grease on the syringe joint. A check for gross leaks is performed by pulling air into the syringe, inserting the needle into a rubber stopper, compressing the syringe plunger to about half of the original volume, and noting whether the plunger returns to its original volume mark when released. Poorly fitting needles or syringe parts should be discarded. Needle tips are easily blunted and bent over in the course of laboratory work, so a small, fine-grained whetstone should be kept handy to resharpen the point. A needle tip that is curled back or has a burr on the point is especially damaging to septa. A sharp needle not only makes insertion easier but also reduces leakage through the punctured septum. The so-called noncoring or deflecting tip, which has a side opening or a 12° beveled tip, is preferred over other types of needles because it does the least damage to septa. A flat-cut point should be avoided. When working with unstable or reactive substances, it is important to flush a needle and syringe with pure solvent immediately after use. Stiff metal cleaning wires are often provided with new needles and are useful for removing deposits from the interior of the needle.

The outside diameter of the needle is generally specified in the United States by its wire gauge. Table 1.2 presents the metric equivalent of common wire gauges and the sizes of matching syringes. The latter are provided as a convenient guide and not as a set of rules. Teflon needles are convenient for transferring highly corrosive materials; however, these are not general substitutes for metal syringes because they cannot penetrate septa. A range of special syringe fittings is available, such as crosses, filters, T's, and stopcocks.[6] The small stopcocks or valves find use in retaining liquids in syringes. These are especially useful in conjunction with large-volume syringes or when handling highly pyrophoric materials, such as aluminum alkyls. A setup including a syringe stopcock along with a syringe holder has already been illustrated (Fig. 1.9). Another application of these items is the adapter for the removal of aluminum alkyls and other reactive liquids from a small cylinder described in the following example.

Table 1.2. Needle Sizes and Matching syringes

Gauge	o.d. (mm)	Approximate Volume of Matching Syringe
25	0.46	Microliters
23	0.57	0.2–2 mL
20	0.81	1–5 mL
18	1.02	5–50 mL
14	1.63	50–100 mL

[6]The construction of a large syringe for dispensing gases is described by G. W. Kramer, *J. Chem. Educ.*, 50, 227 (1973). Syringe T's, stopcocks, and needle stock for cannulae are available from Aldrich Chemical Co., P.O. Box 355, Milwaukee, WI 53201.

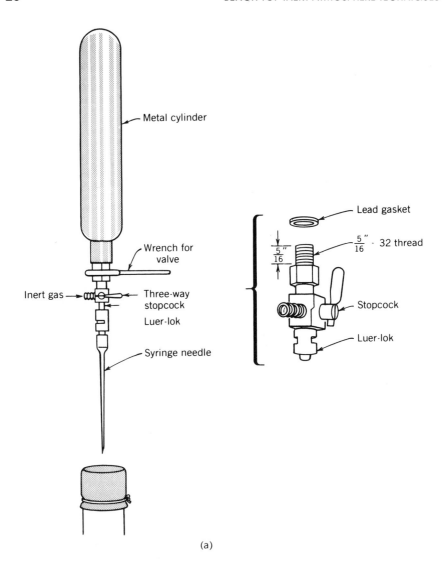

(a)

E. Example: Withdrawal of Highly Reactive Liquids from Metal Tanks.

Highly reactive liquids such as aluminum alkyls can be withdrawn from a low–pressure cylinder using the apparatus shown in Fig. 1.10*a*. **CAUTION: Highly reactive liquids should be handled in a hood which contains a minimum of flammable material or chemicals.** After attaching the syringe apparatus to the cylinder, the needle is purged with inert gas for several minutes. While purging, the needle is inserted through the septum into the reaction flask.

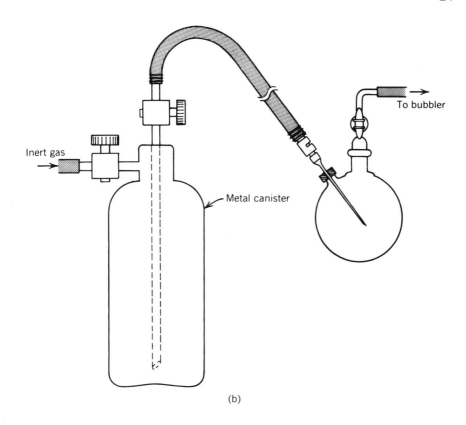

Inert gas

Metal canister

To bubbler

(b)

Fig. 1.10.   Transfer of air-sensitive liquids from metal cylinders. (*a*) A three-way metal stopcock constructed for use with syringes may be modified by the addition of a threaded fitting so that it mates with a lecture bottle. The side inlet is used to purge the apparatus initially and to blow liquid out of the syringe needle at the completion of the transfer. The cylinder is put under a low pressure of inert gas before the transfer is begun. (*b*) Sometimes air-sensitive liquids are sold in siphon-type cylinders, and the liquid from such a cylinder can be dispensed as shown here. As with the previous example, inert gas is used to force the liquid out of the cylinder.

The three-way valve is then switched to allow liquid dispensing. When the desired quantity of liquid has been removed from the cylinder, the cylinder is closed and the three-way valve switched back to inert gas, so that the needle is purged by inert gas before it is removed from the reaction flask.

An apparatus for the withdrawal of reactive liquids from metal cylinders having a siphon tube is shown in Fig. 1.10*b*. This setup is operated in much the same way as the syringe apparatus, the main difference being the use of inert gas to force the liquid out of the cylinder containing the siphon tube.

F. Liquid Transfer Cannulae. The transfer of liquids between two flasks may be accomplished with a cannula alone (Fig. 1.11). These cannulae usually are constructed from syringe needle stock which is commercially available in convenient lengths.[6] In addition to simple liquid transfer, special filtration cannulae can be constructed. In a recent design by M. L. H. Green, a stainless-steel cannula is cemented with epoxy resin to a short length of heavy-walled glass capillary tubing which has a slightly flared and fire-polished opening (Fig. 1.12a). This opening is covered by hardened filter paper which is neatly folded back on the glass fitting and wired in place, as illustrated in Fig. 1.12b. A sleeve-type septum stopper is pushed onto this filter apparatus, with the stopper end toward the filter. This assembly may then be attached to a receiver and flask as illustrated in Fig. 1.11. Filtration with more conventional equipment is illustrated in Fig. 1.13.

In comparison with syringe techniques, the use of cannulae provides better air exclusion and greater convenience for the transfer of large volumes of solution. However, syringe techniques are much better suited for the quantitative dispensing of liquids, as demonstrated in the following example.

G. Example: Quantitative Dispensing of Liquids. Moderate accuracy in liquid-transfer operations can be achieved using syringe graduations. Air should first be expelled from the filled syringe by pointing the needle up and squirting out all bubbles. Higher accuracy can be achieved by weighing the dispensed liquid. The syringe is filled and the needle tip is stoppered by insertion into a small rubber stopper. This assembly is weighed, the liquid dispensed, and the syringe plus rubber stopper is reweighed. If the syringe is not going to be re-

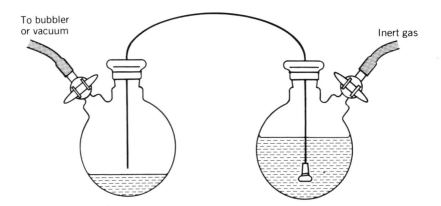

Fig. 1.11. Solution transfer using a stainless steel cannula. The solution is forced through the needle by means of the pressure differential created by opening the inert gas inlet, which pressurizes the right-hand flask. The receiving flask is either maintained at 1 atm by opening the sidearm to a bubbler, or is maintained at slightly less than 1 atm by opening the sidearm briefly to a vacuum source.

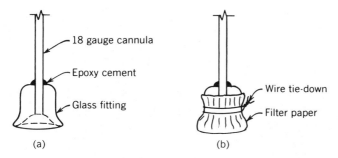

Fig. 1.12. Assembly of a filter cannula or "Green filter." (*a*) A piece of glass capillary tubing, which has been blown open and fire-polished to form a lip on one end, is attached to the cannula with epoxy cement. (*b*) Filter paper is folded over the glass capillary and secured with wire.

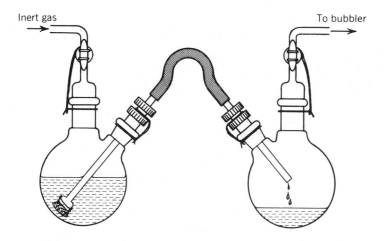

Fig. 1.13. Variation on the cannula technique. The glass tube with fritted filter and the glass delivery tubes are attached to the glass standard taper joints by means of plastic adaptors (e.g., Kontes K-179800). The flexible plastic tubing connecting the two parts should be impervious to solvents. Teflon tubing is best and, to prevent contamination of the filtrate, Tygon tubing should be avoided when organic solvents are used.

used immediately, it should be rinsed with degassed solvent immediately after use, to avoid the buildup of deposits of decomposition products in the syringe and needle. However, if it is to be re-used soon with the same solution, the stopper may be left on the needle tip and both rinsing and inert-gas flush may be omitted before the syringe is filled again.

H. Spectroscopic Measurements. The sampling of air-sensitive liquids for NMR and IR spectroscopy is often most conveniently performed by syringe. Small rubber septa are available which fit 5-mm and other small NMR tubes and also fit the Luer inlets on IR cells. The NMR tube may be purged by pump–and–fill operations, whereas IR cells are easily flushed by inert gas. The filling of the IR cell is accomplished by inserting the syringe needle into the septum on the cell inlet, and the outlet septum is pierced with a small needle. An excess of liquid is flushed through the cell to exclude gas bubbles. A somewhat less air-free operation is achieved by flushing the cell and then attaching a syringe containing the liquid by joining the Luer fittings on the syringe and cell. An empty syringe is attached to the cell exit to serve as a reservoir for excess liquid which is forced through the cell. Once the cell is filled, the inlet and outlet are capped with Teflon stoppers. Conventional visible–UV cells can also be fitted with syringe caps for air-free spectroscopy. A visible–UV cell that has been used for a wide variety of air-sensitive inorganic and biological compounds is illustrated in Fig. 1.14. More sophisticated spectroscopic cells, some of which are suitable for use in conjunction with syringe techniques, are described in Chapter 9.

1.4 QUANTITATIVE GAS MANIPULATION

A. Dispensing Gases. The dispensing of gas to a solution can be accomplished in several ways. For example, a solution may be saturated by bubbling

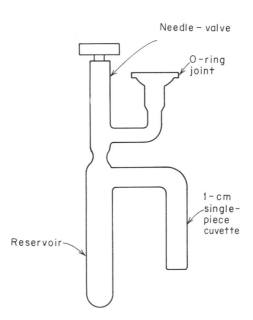

Fig. 1.14. Evacuable cell for UV-visible spectra. This cell is evacuated and then flushed through an inert-gas inlet attached to the O-ring joint. The Teflon stem of the valve is removed under flush, solutions are introduced through this opening, and then tipped over into the cuvette after the valve stem has been replaced. Alternatively, condensable solvents and solutes may be condensed into the reservoir on a vacuum line.

gas through it from a syringe needle attached to a tube and low pressure gas source (Fig. 1.15).

The quantitative delivery of a gas is somewhat more involved. While it is true that a vacuum system often permits the most satisfactory method of quantitative gas handling, some adequate procedures exist which are based on simple, conventional inert-atmosphere apparatus. A large gastight syringe is one of the most obvious devices for the measurement of an aliquot of gas.[6] A sample of gas may be drawn into the syringe from a stream of the gas connected to an exit bubbler. Another method for dispensing given quantities of gas is a gas buret, similar to one described later in connection with the measurement of evolved gases.[7] Still another simple scheme is to flush a bulb of known volume with the desired gas, and then bubble this gas into the reaction flask by means of a slow inert-gas stream (Fig 1.16). Condensable gases may be trapped, measured by volume of the liquid, and dispensed by controlled warming of the gas. A suitable apparatus for this purpose is illustrated in Fig 1.17. Gases which are condensable at Dry Ice temperature, $-78°C$, can be handled by this technique.

Finally, it is often possible to generate quantitatively the desired gas from measured amounts of liquid or solid reagents. The generated gas is then quantitatively transferred by inert-gas flush into a reaction mixture. The automatic gasimeter developed by C. S. Brown and H. C. Brown is useful when it is necessary to know the quantity of gas consumed in the course of a reaction.[8] This apparatus can be applied to the generation of a variety of gases, such as H_2,[8]

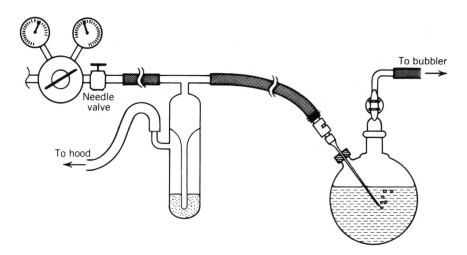

Fig. 1.15. Dispensing gas directly from a tank. The needle valve is used to control the gas flow into the solution, and a mercury-filled bubbler prevents excessive pressure buildup.

[7]T. N. Sorrell and M. R. Malachowski, *Inorg. Chem.*, 22, 1883 (1983), provide a recent description of gas uptake experiments with a gas buret.

[8]C. A. Brown and H. C. Brown, *J. Org. Chem.*, 31, 3989 (1966).

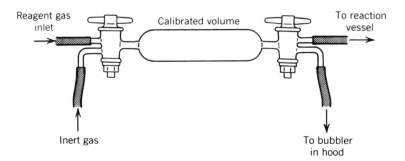

Fig. 1.16. Gas-sampling tube. The tube is flushed with the reagent gas by opening to the reagent gas inlet and venting to a mineral oil bubbler. After several minutes, the flush is terminated by first closing the inlet stopcock, then the outlet stopcock. (This ensures that the pressure inside the tube equals the laboratory atmospheric pressure.) The laboratory temperature and pressure are then measured for use in calculating the amount of gas dispensed. Finally, the known quantity of reagent gas is dispensed to the reaction vessel by opening the inlet stopcock to inert gas, the outlet stopcock to the reaction vessel, and flushing the tube with inert gas for several minutes.

CO,[9] hydrogen halides,[10,11] O_2,[12] ethylene,[11] and SO_2,[11] but not gases which react with mercury. By means of a clever liquid inlet device, gas is produced in the generator in response to the amount consumed in an attached reactor. This liquid inlet valve (Fig. 1.18), consists of a tube partially filled with mercury. The tube has provision for the inlet of liquid from a needle attached to a buret and for the controlled flow of this liquid into the reaction vessel via several small holes above the mercury level. Liquid from this inlet valve drops into the gas generator, and the evolved gas is conducted from there to a reaction flask. The flow of liquid can be initially adjusted by the depth to which the needle projects into the mercury pool of the valve. When the system is properly balanced, the pressure drop caused by gas consumption in the reactor pulls the reagent for gas generation in through the liquid control valve. As the pressure builds up, this flow of reagent stops, only to resume again when the reaction consumes some of the generated gas. When the quantitative measurement of gas consumption is required, the exit bubbler serves only as a safety device and the system must be set up so that there is no gas expelled through the bubbler during the course of the reaction.

B. Measurement of Evolved Gases. The measurement of evolved gases provides a ready means of conducting certain analyses and following the course

[9]M. W. Rathke and H. C. Brown, *J. Am. Chem. Soc.*, 88, 2606 (1966).

[10]H. C. Brown and N.-H. Rei, *J. Org. Chem.*, 31, 1090 (1966).

[11]G. W. Kramer, A. B. Levy, and M. M. Midland, in H. C. Brown, *Organic Syntheses via Boranes*, New York; Wiley, 1973, p. 218.

[12]H. C. Brown, M. M. Midland, and G. W. Kabalka, *J. Am. Chem. Soc.*, 93, 1024 (1971).

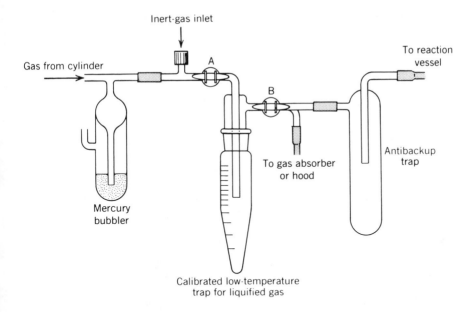

Fig. 1.17. Apparatus for dispensing known quantities of condensable gases. The trap may be calibrated by using water (for large volumes) or mercury (for small volumes). A mercury bubbler is included to prevent blowing the apparatus apart.

After assembling the apparatus, the entire system is thoroughly purged with inert gas. Stopcock B is then turned to route gas to a hood or gas absorber, the inert-gas flow is terminated, and a slow flow of the desired gas is admitted from a cylinder equipped with a pressure regulator and needle valve for flow control. A Dewar containing refrigerant capable of liquifying the gas (see Table 5.1) is then raised around the trap. After the desired volume of liquid is collected, the reagent gas flow is terminated, stopcock A is closed, and stopcock B is turned to direct gas to the reaction vessel. The Dewar is then lowered from the trap, allowing the collected gas to boil off. The rate of vaporization may by controlled by raising and lowering the Dewar. When the trap appears to have been emptied, a flush of inert gas may be used to flush out remaining reagent vapor.

of a reaction. For example, the analysis of active hydrides and alkyl compounds can often be conducted by quantitative hydrolysis to produce hydrogen or alkane. A simple gas buret system, illustrated in Fig. 1.19, is useful for these types of measurements. It also may be used to dispense specific quantities of gas to an inert-gas stream for introduction into a reaction mixture. The procedure for using this gas buret apparatus for the analysis of active hydrides or alkyl compounds by hydrolysis is described in the following example.

C. Example: Analysis of Active Hydrides or Alkyl Compounds Using a Gas Buret.

This procedure utilizes the gas buret illustrated in Fig. 1.19 to analyze active hydrides or alkyls. First, the water levels in the reservoir and buret sides are adjusted to equal height and an initial buret reading is taken. A sample of known weight is introduced into the hydrolysis flask using a syringe. After gas

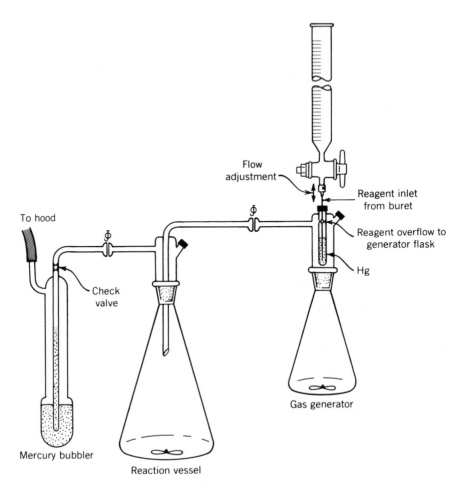

Fig. 1.18. Brown[2] gas generator and apparatus for reactions with gases. Reagents are mixed in the gas generator and the resulting gases are passed into the reaction vessel. The mercury bubbler on the far left has a float check valve which prevents mercury and air from being sucked back into the reaction vessel. The key to this apparatus is the automatic liquid inlet valve, shown in detail on the right. A drop in gas pressure within the apparatus pulls liquid from the buret through the syringe needle and the mercury. The rate of delivery of reagent is adjusted by raising or lowering the buret and attached syringe needle.

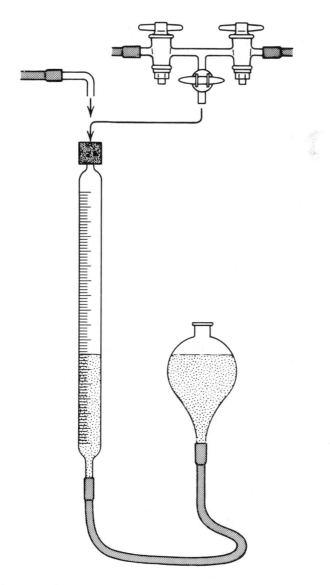

Fig. 1.19. A gas buret for the measurement or delivery of gases. When the elbow (on left) is attached to the top of the buret, gas is collected and its volume is measured after equalizing the mercury levels. For example, a hydrolysis flask similar to the reaction flask in Fig. 1.18 may be attached to this elbow. When the stopcock assembly is used (at the right), the buret can be filled with gas and a measured volume of this gas can be injected into a stream of flowing inert gas.

evolution has ceased, the water levels are again matched, and both a buret and barometric pressure measurement are taken. A correction must be made for the vapor pressure of the solvent used in the system, especially if the solvent is very volatile.

1.5 SCHLENK TECHNIQUES

A. Basic Apparatus. The original Schlenk tube consisted of a long tube with a sidearm near the top which permitted the removal of air by evacuation and filling with inert gas. A complete methodology for work with air-sensitive liquids has evolved from the original Schlenk tube. The Schlenk tube described here has a teardrop shape to facilitate the stirring of reaction mixtures and the vacuum evaporation of solvents, while still maintaining ease of pouring and scraping solids from the Schlenk tube. The currently popular Schlenk ware permits the transfer of liquids or solids by pouring within a closed apparatus or under an inert-gas flush. Since it is possible to perform fairly complex solids transfer with this type of apparatus, the need for glove box operations is minimized. Some of the basic items of equipment include a Schlenk tube (which is used as a reaction flask), a fritted funnel, a dropping funnel, and a solids container (Fig. 1.20). Most Schlenk ware pieces are fitted with a sidearm through which the apparatus can be evacuated and filled with inert gas. Some basic operations using several of these pieces are described below.

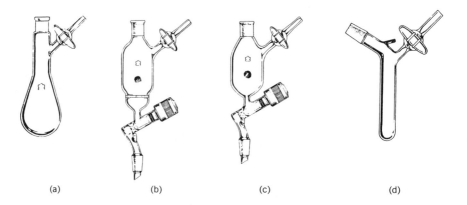

(a)   (b)   (c)   (d)

Fig. 1.20. Basic pieces of inert-atmosphere Schlenk-type apparatus. (*a*) Schlenk tube. This is useful as a reaction vessel and filtrate receiver. The pear shape of the vessel facilitates stirring the contents with a stirbar and magnetic stirrer, and removing solids. (*b*) Fritte or fritted funnel. This is used for filtrations. The Teflon valve is useful when contamination by stopcock grease is objectionable. Double-ended frittes (Fig. 1.5) are also useful. (*c*) Dropping funnel. As in the case of the fritted funnel, the Teflon valve is used to avoid contamination of the solution by stopcock grease. (*d*) Solids container. This is used for adding, receiving, and storing air-sensitive solids. (Reproduced by permission of the copyright owner, Kontes Glass Co.)

B. Example: Purging Schlenk Ware. In general, a high vacuum is not required for the pump–and–fill type of purge used in Schlenk techniques. Several cycles of pumping and filling with inert gas are sufficient to reduce the oxygen and other atmospheric gases to very low levels. For example, suppose the system is evacuated to 2 torr. Applying Eq. (1), after the second pumping and filling the pressure of residual air is $2 \times 2/760$ torr, and after the third it is $2 \times (2/760)^2$ torr, or approximately 10^{-5} torr. Because of the frequency with which these pump–and–fill cycles are performed, it is useful to employ a manifold for the distribution of vacuum and inert gas, such as the unit illustrated in Fig. 1.2.

C. Example: Joining Two Pieces of Schlenk Ware. Another basic operation is the joining of apparatus under an inert-gas flush. For example, the transfer of a reactive solid from a solids container to a Schlenk tube requires the tube to be stoppered and carried through several purge cycles. Then, with both pieces open to a nitrogen source, the cap is removed from the solids container, the stopper is removed from the Schlenk tube, and the two pieces are quickly joined while air is excluded by the flow of inert gas out of the two pieces of apparatus, as previously illustrated in Fig. 1.5. These operations require that the pressure in the system be only slightly above atmospheric pressure so that the closed apparatus is not blown apart. A two stage regulator with good control in the low pressure range is essential, and additional protection against overpressurization can be achieved by the pressure release bubbler illustrated in Fig. 1.2. To maintain integrity of the apparatus, springs, clips, or rubber bands are used to secure each joint, and stopcocks are held in place with sturdy retainers.[2] A similar sequence of operations is used for changing the configuration of all types of Schlenk apparatus during the course of a synthesis or purification.

The apparatus for several other common Schlenk operations are shown in Figure 1.21. These operations include filtration (Fig. 1.21a), transferring solids from a filter to a solids container (Fig. 1.21b), and transferring solids from a solids container to an ampule which will be sealed off (Fig. 1.21c). Alternatives for adding solids to ampules are illustrated in Fig. 1.22.

D. Quantitative Addition of Liquids and Solids. The quantitative measurement of air-sensitive solids and liquids can be accomplished in a number of ways. Chapter 2 will cover glove box operations which are suited for the measurement of solids. The quantitative delivery of solutions is readily accomplished by the syringe techniques, which have been discussed earlier in this chapter (Section 1.3.G). In addition, liquids can be added from graduated Schlenk-type vessels which are fitted with a sidearm for flushing. Microweighing tubes (Fig. 1.23) are useful for the delivery of small known weights of compounds to Schlenk apparatus. The solid is introduced into the tube in an inert-atmosphere glove box and the tube and contents are then weighed. With a flush of inert-gas issuing from the Schlenk apparatus, the weighing tube is quickly uncapped, inserted into the Schlenk apparatus, and shaken to deliver its contents. The tube is removed and quickly recapped and reweighed to give the tare weight. Sodium- and

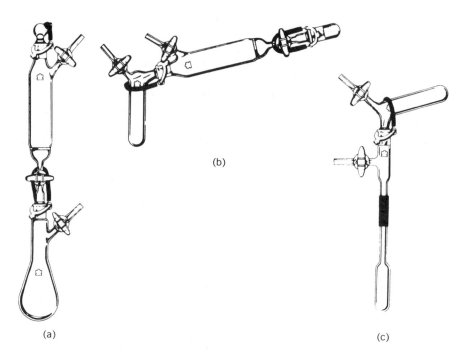

Fig. 1.21. Typical assemblies of Schlenk ware. Commercial apparatus is illustrated; similar apparatus with O-ring joints in place of standard taper joints is preferred by the authors. (*a*) Filter and Schlenk flask receiver; (*b*) pouring solid from a filter into a solids storage tube; (*c*) pouring solid from the solids storage tube into an ampule, which can be sealed off. All apparatus is purged by pump-and-fill operations. Whenever the apparatus is opened or being joined, a purge of inert gas from the sidearm is used to exclude air. (Reproduced by permission of the copyright owner Kontes Glass Co.)

potassium-filled tubes can be cut, measured, and dropped into a reaction vessel rapidly to dispense a known weight of these metals. Another method for dispensing weighed amounts of alkali metals in thin breakable bulbs is described in Section 9.4A. Solids also can be dispensed from sealed ampules, which are broken open under an inert atmosphere, (Fig. 1.24) and quickly poured into Schlenk apparatus under an inert-gas flush. A full ampule can be weighed and the parts tared after the solid is added, or a method described in the next paragraph may be used to determine the weight of sample.

A known amount of material may be sealed in an ampule by weighing the ampule, adding solid, sealing, weighing the parts, and determining the weight of the contents by difference. Filling an ampule is accomplished with the apparatus illustrated in Fig. 1.21 or 1.22, or in a glove box. In order to achieve a good seal, the system must be under vacuum or at most at atmospheric pressure. When the ampule is attached to a Schlenk apparatus, the system can be partially evacuated and the seal then made, or a remote section of the Schlenk line can be momentarily opened up to vent the system while the seal is being made. Similarly, an ampule which has been filled in an inert-atmosphere glove box may be

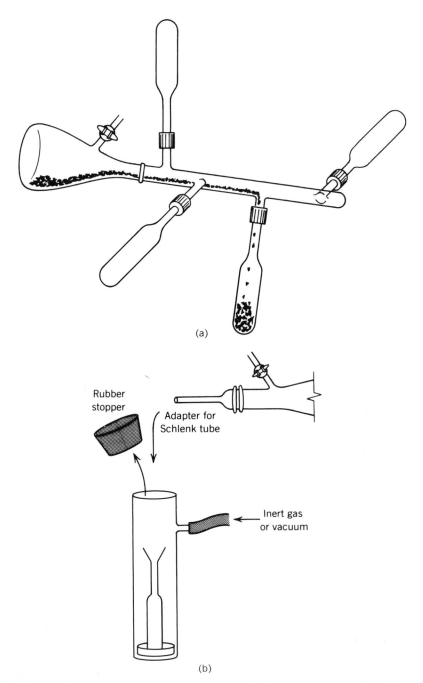

(a)

Rubber
stopper

Adapter for
Schlenk tube

Inert gas
or vacuum

(b)

Fig. 1.22. Solids transfer under an inert atmosphere. (*a*) Solids are poured down this solids manifold into ampules, which are then sealed off. (*b*) An ampule is purged in the evacuable cylinder, and under an inert-gas purge the stopper on the cylinder is removed and solids are transferred from the Schlenk tube adapter into the vial. The latter is quickly removed from the cylinder and sealed off.

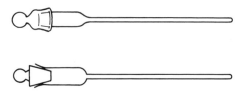

Fig. 1.23. Weighing pigs with caps.

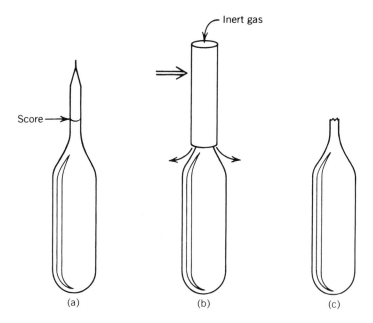

Fig. 1.24. Opening a sealed ampule. (*a*) The neck of the ampule is carefully scored using a glass knife. (*b*) A metal tube of larger diameter than the ampule neck is placed over the neck. Inert gas flowing through the metal tube will minimize exposure of the ampule contents to air once the neck is broken. A quick jerk is then applied to the metal tube, which snaps the neck off (*c*) at the score.

This operation is easily modified for use in a glove box. The score is administered to the ampule neck before transferring the ampule to the glove box. The ampule is clamped to a stand in the glove box and the neck snapped using a cork borer. This minimizes the chances of glass fragments cutting a glove.

stoppered with a sleeve septum cap which is pierced with a small needle just before the ampule is sealed. A small residue of the compound remaining in the neck of the ampule may cause defects in the seal which allow the entrance of air. To minimize this problem, the glass is first collapsed in the usual sealing operation, and then, as the flame is kept on the collapsed portion, the parts are pulled apart with a twisting motion. Less than one full revolution is generally adequate to close the threadlike channel which would otherwise develop. Details on this and other glass blowing operations are given in Appendix II. When an evacuated ampule is broken open, there is an inrush of air. To minimize exposure of the contents to air, the ampule may be broken open as illustrated in Fig. 1.24 and the contents quickly poured into Schlenk apparatus which is being flushed with inert gas.

E. Separations. The basic filtration operation has been previously illustrated. Low-temperature filtrations for the collection of thermally sensitive compounds or products which are soluble at higher temperatures may be performed with an H-type Schlenk tube, Fig. 1.30, in a large low-temperature bath. The jacketed, fritted funnel illustrated in Fig. 1.25 also permits low-temperature filtration.

Crystallization is performed in Schlenk apparatus under an inert atmosphere using familiar techniques: partial removal of solvent by evacuation, addition of a solvent in which the solute has reduced solubility, and cooling. These methods are often combined; for example, a second, poorer solvent, chosen so that its vapor pressure is less than that of the more powerful, original solvent, may be added. Partial evacuation will then concentrate the solution and change the solvent composition to favor crystallization. The complication which air-sensitive

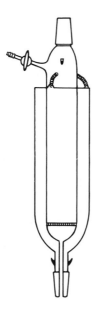

Fig. 1.25. Low-temperature filtration apparatus. This is utilized the same way as the fritte pictured in Fig. 1.20b. The outer jacket is filled with a Dry Ice/acetone solution when filtering heat-sensitive compounds, liquid ammonia solutions, etc. [Reproduced by permission of the copyright owner, The American Chemical Society, from J. E. Ellis, K. L. Fjare, and T. G. Hayes, *J. Am. Chem. Soc.* 103, 6100 (1981).]

materials present is that impurities created by partial reaction with adventitious air or moisture may make a product difficult to crystallize. If these impurities are insoluble, they can be filtered off before the crystallization operation; but soluble impurities present a more serious problem. The layering of a poor solvent on a solution of a compound is often used to grow crystals for crystal structure determinations. In this case, the crystals grow at the interface between the two liquid phases. If the two liquid phases have significantly different densities, the more dense should be placed in the bottom of the tube. It is best for the initial solution to be far from the saturation concentration of the compound. Tubes fitted with sleeve-type serum caps can be used for moderately sensitive compounds, and a right-angle Teflon valve can be used to close the crystal growing tube for more sensitive compounds. When suitable crystals have grown, the

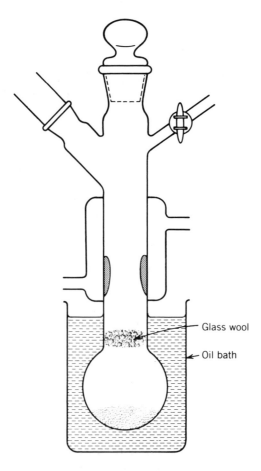

Fig. 1.26. Sublimation apparatus. Solids are introduced under inert-gas flush, and a loose plug of glass wool is introduced. The oil bath should extend above the glass wool. High vacuum may be applied through the standard taper joint. A lower vacuum may be applied through the sidearm.

liquids may be carefully removed by syringe and crystals shaken from the tube in a glove box. Sometimes it is necessary to break the tube to gain access to the crystals.

Sublimation is readily performed in Schlenk apparatus of the type illustrated in Fig. 1.26. The sublimate, which collects on the upper walls, may be collected by positioning the apparatus horizontally and scraping the product into a receiver attached to the side joint while an inert-gas flush is maintained on the equipment.

Column chromatography under an inert atmosphere is made possible by the apparatus illustrated in Fig. 1.27. Most common adsorbents contain large quantities of moisture and air, the elimination of which is difficult. For example, it is generally impractical to eliminate all of the surface hydroxyl groups from alumina. Also, the absorbency of inorganic solids such as silica gel and alumina is often a strong function of adsorbed moisture. A convenient procedure for degassing and partial moisture removal of the adsorbent is to warm the solid in a Schlenk tube while carrying out a series of pump–and–fill operations. (To avoid particles of adsorbate flying throught the Schlenk manifold, a small plug of glass wool is placed in the neck of the flask, and vacuum is applied slowly while the vessel is swirled.) Under a blanket of inert gas, air-free solvent can then be added with a dropping funnel and the slurry poured into the column (generally, a transfer cross is necessary so that the slurry can be washed into the column by means of solvent introduced by syringe). Alternatively, the dry adsorbent in the column

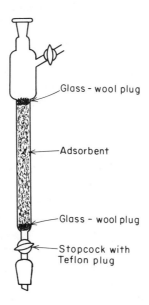

FIG. 1.27. Column for adsorption chromatography. Typical size: 14/35 ℑ joints, 45-mm-diameter upper section, 20-mm-diameter column, and a 500-mm column length. O-joints and Teflon-in-glass valves provide greater solvent resistance than the greased fittings illustrated here.

can be deaerated by pump–and–fill operations, and solvent added later. Occluded gas is then eliminated under an inert-gas flush by stirring the column with a long wire or rocking the slurry back and forth in the closed column. One sidearm on the column is used to provide a nitrogen flush when the system is open to the air, and another may be added for the introduction of small samples by syringe.

F. Molecular Weight Determinations. Molecular weights of slightly air-sensitive compounds can be determined in standard commercial vapor pressure osmometers equipped with an inert-gas purging line. Argon is preferred to nitrogen for purging this apparatus; the primary leakage paths are around the top of the apparatus and thus a heavy gas is desirable. Standard boiling point elevation apparatus can be equipped with an inert-gas supply to provide very good exclusion of air and thus permit the collection of molecular weight data on highly air-sensitive compounds.[13]

The measurement of molecular weights by freezing point depression is conveniently accomplished on air-sensitive compounds by means of the apparatus illustrated in Fig. 1.28. In this design, the sample is introduced to the cell from a weighing tube or syringe under a nitrogen flush, or it may be introduced in an inert-atmosphere glove box. The microstirring bar in the sidearm of this apparatus is cooled with liquid nitrogen and flicked by means of a magnet into the cell when the solution supercools. The temperature is sensed with a thermistor, and a stripchart recording of the cooling curve may be extrapolated by a linear approximation if the region of supercooling is short, or by more complex functions described in the literature.[14]

Cell designs to satisfy a variety of needs are described in the literature.[15-17]

G. Variations in Schlenk Ware. A large number of modifications of the basic Schlenk ware are in use. The double-ended filter is a commonly used variant of the standard Schlenk filter. The particular unit illustrated in Fig. 1.29 is equipped with O-ring joints rather than standard taper joints. (See Chapter 8 for a discussion of joints and O-ring materials.) Either conventional or modified O-ring joints, such as the Fischer and Porter Co. Solv-Seals, have several advantages over standard taper joints for Schlenk apparatus: these O-joints are more resistant to solvents than the conventional standard taper ware, the O-joints are more readily clamped together, and, unlike standard taper joints, the two halves

[13]F. W. Waker and E. C. Ashby, *J. Chem. Educ.*, 45, 654 (1968).

[14]W. J. Taylor and F. D. Rossini, *J. Res. Natl. Bur. Std.*, 32, 197 (1944).

[15]Enclosed cell for cryoscopy in sulfuric acid, using a Pt resistance thermometer: R. J. Gillespie, J. B. Milne, and R. C. Thompson, *Inorg. Chem.*, 5, 468 (1966).

[16]Enclosed cryoscopy cell with thermistor sensor: T. L. Brown, G. L. Gerteis, D. A. Bafus, and J. A. Ladd, *J. Am. Chem. Soc.*, 86, 2135 (1964).

[17]Enclosed cell with a conventional thermometer: S. U. Choi, W. C. Frith, and H. C. Brown, *J. Am. Chem. Soc.*, 88, 4128 (1966).

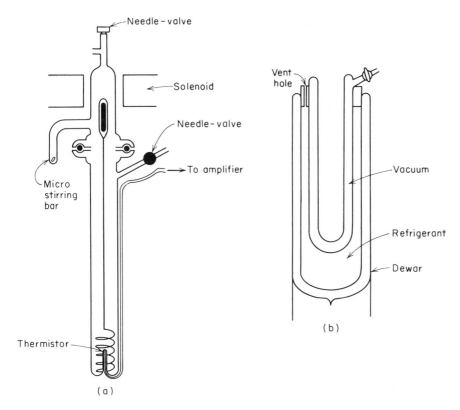

Fig. 1.28. Cryoscopy cell. (*a*) Typical dimensions of the lower section are 260-mm height and 20-mm diameter. The thermistor well is about 27 mm. (*b*) Cooling apparatus for the cryoscopy cell. The cryoscope slips into the jacketed tube. The jacket is evacuated or partially evacuated to control the cooling rate. Electrical readout is achieved with a digital multimeter or commercial electronics for thermistor thermometers.

of these O-ring joints are identical. When moderate quantities of solids are being handled, a fairly complete setup for inert-atmosphere preparations and purifications consists of the double-ended filter in conjunction with Schlenk tubes, solvent storage flasks, solids storage tubes, syringes, and addition funnels.

The H-type Schlenk tube, sometimes called a double Schlenk tube, is yet another type of general-purpose inert-atmosphere apparatus, (Fig. 1.30). This apparatus is basically a combination of two Schlenk tubes and a filter. One advantage of the H-type Schlenk tube is that it forms a very sturdy, leak-tight unit. In addition to the usual pouring of solutions through the filter, it permits the distillation of volatile solvents from one leg to the other. This last feature can be used to wash and dry a precipitate. The H-type Schlenk tube also is more satisfactory than conventional Schlenk ware for low-temperature filtration. Weighed against these advantages, the H design is unnecessarily bulky for some operations and it is somewhat more difficult to use for large quantities of solids. As with the con-

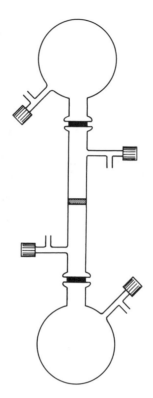

Fig. 1.29. Double-ended filter. This type of apparatus is a good alternative to the more standard Schlenk filter. The version illustrated is based on greaseless joints and valves.

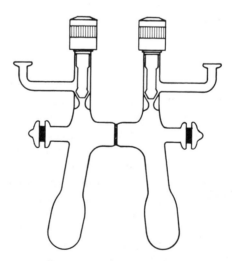

Fig. 1.30. H-Schlenk tube. In addition to replacing the standard Schlenk filter for routine operations, this apparatus permits low-temperature filtration if a large cooling bath is used. See Chapter 5 for the use of the H-tube in the Wayda-Dye vacuum apparatus.

ventional designs, a complete set of common operations for synthesis and/or purification can be designed around the H-tube.

1.6 CAPPABLE PRESSURE REACTORS

A complete methodology for the manipulation and reaction of air-sensitive solutions has evolved around cappable glass pressure bottles. Soft-drink bottles are sometimes used (hence these procedures are sometimes referred to as "pop bottle techniques"); however, heavy-walled borosilicate glass pressure reaction vessels are superior. In contrast to the modified standard taper ware discussed above, this pressure apparatus offers advantages where modest pressures are necessary and where the centrifugation of precipitates is preferable to filtration. These techniques are especially popular in the preparative-scale study of catalytic reactions of small molecules, such as olefin polymerization. The pressure bottle is fitted with a cap containing two 1/8 in. holes and a rubber liner, which is secured by means of a hand-operated bottle capper (Fig. 1.31).[18]

Depending on their sensitivity, solids are loaded into pressure reaction bottles in the air or in an inert-atmosphere glove box. The bottle is capped and may be carried through several purge cycles by means of a hypodermic needle attached

Rubber
liner

Fig. 1.31.   Heavy-walled centrifuge-tube, rubber cap liner, and cap. ("Pop bottle" reactor.) Tubes with volumes ranging from 20 to 250 mL are available from Lab Glass, North West Boulevard, Vineland, NJ 08360.

[18]Cappable reaction bottles, caps, and the rubber liner septa are available from Lab Glass, Inc., Vineland, NJ and from Aldrich Chemical Co., Milwaukee, WI. Bottle cappers are available from large hardware stores.

to a dual inert-gas and vacuum manifold (Fig. 1.2). Ordinarily, the reactor is left under a small positive pressure (900–1000 torr, 3–5 psig), which minimizes the entrance of air and facilitates the removal of liquids by means of a syringe. Reactions which evolve large quantities of gas may require occasional venting through a hypodermic needle. Pressures in excess of 3,300 torr (50 lb/in.^{2}) should be avoided. The solution may be agitated by means of a shaking machine or by means of a magnetic stirring bar which was included with the initial charge. Since the bottles are easily thermostated and samples are readily removed by syringe, the cappable reactors have proven useful for kinetic investigations of air-sensitive solutions. These reactors are available in shapes which fit large centrifuges, and are therefore quite useful for separating solutions from solids which do not filter well. Reactive solids transfer operations are readily performed in a glove box because these sealed reaction vessels are easily transferred through a vacuum lock.

1.7 HOT TUBE AND SEALED TUBE TECHNIQUES

A. Hot Tube Reactors. Many simple metal halides, nitrides, and the like are conveniently synthesized, and sometimes purified, at elevated temperatures in an apparatus designed to fit in a tube furnace. The general strategy is to contain one or more of the starting materials within the heated tube. Often a reactive gas is passed through the tube. In many instances the reactive gas is diluted with an inert gas, and the inert gas is used when purifying the product by sublimation. One example, given in Fig. 1.32, is the preparation of a nonvolatile product, magnesium nitride, from magnesium metal and nitrogen gas. When the reaction is complete, a back-flush of nitrogen is maintained from the attached Schlenk tube as the inlet joint is removed. A long, stiff wire is then used to push the boats containing the magnesium nitride into the attached Schlenk tube. The second example in this same figure illustrates the procedures for a volatile product, gallium chloride.[19] In this case the reaction is performed in an HCl stream and the product is subsequently sublimed into a series of bulbs, which can then be sealed off. The stream of inert gas has to be adjusted carefully during the sublimation, so that the product is carried into the bulbs, but the inert-gas stream should not be so rapid that the sublimate is swept out of the apparatus. In the course of a typical hot tube reaction, large volumes of gas are exposed to highly reactive materials in the reactor, thus exposing the materials to the cumulative amount of impurities in the gas. It is necessary, therefore, to use pure gases or to purify all gases by appropriate dessicants and/or oxygen scavengers. Chapter 3 may be consulted for details on inert-gas purification.

---

[19]W. C. Johnson and C. A. Haskew, *Inorg. Synth.*, 1, 26 (1939), present details of the gallium chloride synthesis.

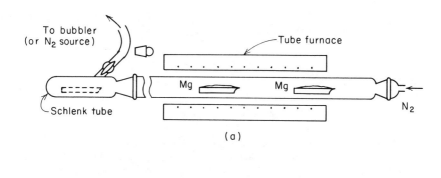

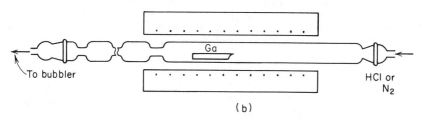

Fig. 1.32. Reactions in heated tubes. (*a*) Apparatus for the preparation and collection of magnesium nitride. Highly purified nitrogen is used, and only the product from the second boat is collected. Sample collection is performed after the tube cools by reversing the nitrogen flow through the apparatus, inserting a long, stiff wire from the right end, and pushing the boat into the Schlenk tube. (*b*) Preparation of a volatile halide. Very dry gas is used, and its flow rate is carefully adjusted so that the halide neither diffuses back toward the gas source nor is swept past the first bulb. After the reaction is complete, the halide may be resublimed in a gas stream and then sealed off in one or more of the bulbs.

B. Sealed Tube Preparations.   Sealed tube reactions and purifications afford good exclusion of the atmosphere and permit operations at moderate pressures. The latter capability makes it possible to work with highly volatile solvents, such as liquid ammonia, at room temperature. Since the sealed tube is often loaded on a high-vacuum line, the details on these techniques are given in Chapter 9.

Another type of sealed tube process is employed for chemical vapor transport used to synthesize and purify many inorganics, such as metal disulfides. In the most common applications of these techniques a tube with one end sealed is charged with an impure or polycrystalline sample of the desired compound plus a small quantity of a transport reagent. The tube is then evacuated and sealed at the other end. The sealed tube is placed in a furnace capable of maintaining a temperature gradient, with the hotter end corresponding to the charged materials and the cooler end empty. For example, the material to be purified might be TaS_2 with a small amount of I_2 added as transport agent. Reaction of this transport agent with the TaS_2 yields volatile products, which are reconstituted as purified single crystals of TaS_2 at the colder end of the tube.[20,21]

GENERAL REFERENCES

Angelici, R. J. 1977, *Synthesis and Technique in Inorganic Chemistry,* 2nd ed., W. B. Saunders, Philadelphia.

Brauer, G., Ed., *Handbuch der Preparativen Anorganischen Chemie,* 3rd ed., Vol. 1, 1975; Vol. 2, 1978; Vol. 3, 1981. This book contains some general sections on laboratory technique and many examples of the use of inert atmosphere techniques in the individual procedures for the synthesis of inorganic compounds. See especially "Preparative Methoden," p. 1, Vol. 1, by P. W. Schenk, R. Steudel, and G. Brauer; and "Metallorganische Komplexverbindungen," by W. P. Fehlhammer, W. A. Herrmann, and K. Ofele, Vol. 3, p. 1978. The second edition is available in an English translation but is less up to date than the above material.

Brown, H. C., 1973, *Organic Syntheses via Boranes,* Wiley-Interscience, New York. See Chapter 9, p. 191, "Laboratory Operations with Air-Sensitive Substances," by G. W. Kramer, A. B. Levy, and M. M. Midland, which stresses syringe techniques.

Eisch, J. J., and R. B. King, Eds., *Organometallic Syntheses,* Academic Press, New York, Vol. 1, by R. B. King, 1965; Vol. 2, by J. J. Eisch, 1982.

Gysling, H. J., and A. L. Thunberg, 1986 "Techniques for Handling Air Sensitive and Toxic Materials," in B. W. Rossiter, *Physical Methods of Chemistry,* Wiley-Interscience, New York, Vol. 1, 1986, p. 373.

Herzog, S., J. Dehnert, and K. Luhder, 1968, in H. B. Jonassen and A. Weissberger, Eds. *Technique of Inorganic Chemistry,* , Wiley-Interscience, New York, p. 119.

Jolly, W. L., 1970, *The Synthesis and Characterization of Inorganic Compounds,* Prentice-Hall, Englewood Cliffs, N.J.

[20]Greater detail and further examples may be found in H. Schaefer, *Chemical Vapor Transport Reactions*, New York, Academic, 1964.

[21]The synthesis of tungsten oxyhalides by this process is described by J. Tillack, *Inorg. Synth.*, 14, 109 (1973) and details of the TaS_2 purification are given by J. F. Revelli, *Inorg. Synth.*, 19, 35 (1979).

2

INERT-ATMOSPHERE
GLOVE BOXES

This chapter describes the principles of inert-atmosphere glove box operation, the design of glove boxes, and glove box procedures. The beginner in a laboratory with an operating glove box may wish to start by reading Section 2.1, and the concluding section, "Equipment and Operations." Other sections can then be consulted as the need arises. If the purchase or design of a new system is being considered, the whole of this chapter should be useful. The purification of inert gases, discussed in Chapter 3, is also recommended for individuals who wish to gain a thorough understanding of inert-atmosphere glove boxes.

2.1 GENERAL DESIGN AND APPLICATIONS

The inert-atmosphere glove box provides a straightforward means of handling air-sensitive solids and liquids. In its simplest form, it consists of a gas tight box fitted with a window, a pair of gloves, and a gas tight door or transfer port. The entire box is flushed with an inert gas, after which samples may be manipulated in the inert atmosphere.

The glove bag is a simple and inexpensive variant of the dry box. It consists of a large polyethylene bag fitted with a nitrogen inlet and an open end, which may be closed by rolling and clamping. The bag is purged by several cycles of filling with inert gas and collapsing, or by a continuous flush. In its primitive form, manipulations are accomplished in an ordinary bag, but commercially available glove bags with integral polyethylene gloves are much more convenient (Fig. 2.1).[1]

As will be described, one of the primary sources of contamination of the dry box atmosphere is diffusion of air through the gloves. In a few cases this contamination has prompted the use of manipulators which, for simple tasks, need not be intricate. However, the use of manipulators has not met with wide favor, probably because other methods are simpler and cheaper.

The appealing simplicity of glove box manipulations must be weighed against the disadvantages, which are the slowness of operation and the difficulty in maintaining a water- and oxygen-free atmosphere. Work in a dry box can be slow because of the need to pass materials through an air lock and because the gloves and glove ports reduce the experimentalist's dexterity. In addition, there is the problem of maintaining an air-free atmosphere, which is complicated by impurities in the inert-gas supply, leaks in the glove box, and diffusion of moisture and oxygen through the gloves. Solvent vapors or volatile reagents contaminate the atmosphere, and this source of cross-contamination can become serious if several individuals handling quite different materials are using the same dry box. As described later in this chapter, a recirculating gas purifier or a rapid

Fig. 2.1. Polyethylene glove bag. Several sizes are available, including glove bags with two pairs of gloves. (Adapted from Instruments for Research and Industry, Cheltenham, PA, Glove Bag brochure.)

---

[1]Glove bags are available from Instruments for Research and Industry, Cheltenham, PA 19012, and Aldrich Chemical Co., P.O. Box 355, Milwaukee, WI 53201.

flush of inert gas through the box is often employed to minimize impurities in the glove box atmosphere. The most commonly used inert gas is nitrogen; but when the materials being handled form nitrides, helium or argon is used.

Some very active research laboratories in the United States use glove boxes for practically all of their chemical operations, including reactions in solvents. In general, these situations involve one glove box per investigator and special features which allow quick entry and exit from the glove box. By contrast, in many European laboratories, very little, if any, use is made of glove boxes. Instead, various means are devised for handling solids with Schlenk-type equipment. In the authors' laboratories, glove boxes are used for intricate solids transfer operations, but most reactions and the handling of solutions are performed in Schlenk-type apparatus or in a chemical vacuum line. Operations such as weighing reactive solids, making mulls for infrared spectroscopy, and loading capillary tubes for X-ray diffraction are conveniently performed in a glove box or glove bag.

In addition to their use for handling air-sensitive compounds, glove boxes have been used to good advantage in the transfer of radioactive, poisonous, and hazardous biological materials. The strong feature of the glove box for these applications comes from the ease of containment. Because of this characteristic, much of the development of inert-atmosphere glove box design has occurred in conjunction with atomic energy programs.

General-purpose inert-atmosphere glove boxes are available commercially, so it rarely makes sense to construct a system of this type locally.[2] Nevertheless, a knowledge of the construction features is useful for the choice of an appropriate glove box and for the maintenance and modification of an existing inert-atmosphere enclosure. In addition, there are some special types of glove boxes which must be constructed by the user because they are not generally available commercially.

2.2 REPLACEMENT OF AIR BY AN INERT ATMOSPHERE

There are two common methods of replacing air by an inert gas in the main chamber of an inert-atmosphere glove box: evacuation followed by filling with the desired inert gas, or a lengthy purge of the box by the inert gas. The second alternative is generally employed for large boxes, since extremely heavy and well-braced construction is necessary to prevent the collapse of a large evacuated chamber. On a small scale, the structural problems are not so formidable, so a pump-down dry box is attractive if a compact unit is satisfactory. Similarly, a

[2]Manufacturers of inert-atmosphere glove boxes and accessories include Vacuum Atmospheres Co., 4652 W. Rosecrans Ave., Hawthorne, CA 90250 (aluminum or stainless steel construction); Kewaunee Scientific Engineering Co., Adrian, MI 49221 (metal construction); Labconco Corp., 8811 Prospect, Kansas City, MO 64132 (fiberglass-reinforced polyester construction); Manostat Corp., 518 8th Ave., New York, NY 10018 (polymethacrylate construction).

small pumpable transfer lock is generally attached to a large box to facilitate the introduction of apparatus with minimum contamination of the glove box atmosphere.

A. Pump-and-Fill. Providing the apparatus is leak-tight, it is not necessary to attain a high vacuum in a pump–and–fill operation for eliminating air from a transfer lock or main chamber of a glove box. As discussed in Chapter 1, the pump–and–fill process may be repeated until the desired degree of purity is attained. For the case in which a vacuum of f atmospheres is attained on each of n cycles of pumping and filling, the fraction of air A_f remaining is given by Eq. (1):

$$A_f = f^n \tag{1}$$

Thus, a pump which is capable of quickly achieving 1 torr (1.3×10^{-3} atm) will yield a residual air concentration of 1.7 parts of air per million parts of gas after two cycles. It must be stressed, however, that the limitation on the vacuum must not be leakage, because in this case the purity of the gas cannot be improved by repeated pump–and–fill cycles. Therefore, the standards of construction of the apparatus should be appropriate for high vacuum, even though a high vacuum is never achieved. To ensure that the system is leak-tight, the chamber and plumbing to the pump can be subjected to a helium leak tester (Section 7.3D). Similarly, the pump can be checked on a high-vacuum test bench or vacuum line.

Since evacuation is usually the slowest step, some attention should be given to matching the pump to the system. For a chamber of volume V and a pump of pumping speed S, the time required to reduce the pressure from P_1 to P_2 is given by Eq. (2).

$$t = \int_{P_1}^{P_2} \frac{V}{SP}\, dP \tag{2}$$

Since the pumping speed of a mechanical pump does not vary greatly at moderate pressures (e.g., Fig. 6.3), it is reasonable to assume an average constant pumping speed if very low pressures are not involved. Integration then yields Eq.(3).

$$t = \frac{2.3V}{S} \log \frac{P_1}{P_2} \tag{3}$$

As an example, consider the time required to pump a large air lock with a volume of 50 L from a pressure of 760 to 0.5 torr. The free-air capacity of a widely available pump is 140 L/min, but three-fourths of this is a more reasonable average pumping speed for the above pressure range. Insertion of these numbers into Eq. (3) gives a pump-down time of 3.5 min. This value may somewhat underestimate the pumping time, but the calculation illustrates a method for approximately matching the pump to the patience of the experimentalist.

B. Purging. Figure 2.2 illustrates three simple models for flushing air from a chamber by a stream of inert gas: perfect displacement, perfect mixing, and short circuit. The most desirable condition for purging would be perfect displacement, since this involves a gradually moving front of purge gas which pushes all of the unwanted air ahead of it (Figure 2.2a). Under this condition only one chamber volume of inert gas would be required to obtain a perfect purge. In any practical case, perfect displacement cannot be attained, owing to interdiffusion, turbulence, and lack of an ideal geometry for the chamber. After due regard is taken to ensure the proper positioning of the gas inlet and outlet, it appears that interdiffusion is a major limiting factor in the attainment of perfect displacement. Thus, rapid flushing has been found to be more efficient than slow flushing.[3] If, on the other hand, the perfect mixing situation occurred, the entering purge gas would immediately mix with the gas in the chamber and the mixture would exit the chamber at a rate equal to that of the entering purge gas (Figure 2.2b). The differential equation for the latter case may be integrated to yield a simple exponential decay of the fraction of air remaining, f, in terms of the number of chamber volumes of gas passed through the system, n, giving Eq. (4):

$$A_f = e^{-n} \tag{4}$$

In contrast to perfect displacement, Eq. (4) demonstrates that perfect mixing leads to a reduction in the air content by a factor of e^{-1} (0.37) after one chamber volume of purge gas is consumed. Similarly, it may be shown that 8.8 chamber volumes are required to reduce the air content to 150 ppm, and about 10 are necessary to attain 45 ppm. Unless a short circuit occurs because of the inappropriate placement of entrance and exit ports (Fig. 2.2c), the actual situation can be expected to fall between perfect displacement and perfect mixing. Experi-

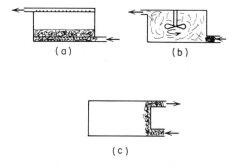

(a) (b)

(c)

Fig. 2.2. Schematic representations of idealized purging conditions. (a) Perfect displacement; (b) perfect mixing; (c) short circuit caused by the introduction of a light gas at the bottom of the chamber.

[3]S. I. Cohen and J. M. Peele, USAEC Report ORNL-CF-55-10-132 (1955).

mental results which illustrate this situation are presented in Fig. 2.3, and a recommended arrangement for the inert-gas inlet and outlet is illustrated in Fig. 2.4*a*.

C. Recommended Conditions. Because of structural considerations, it is generally most satisfactory to use the purging method for large chambers, such as the main chamber of a large dry box. For smaller glove boxes and glove box antechambers, air is conveniently removed by either pump–and–fill operations or purging. If purging is the method chosen for replacing air in the antechamber, it is desirable to include a small manifold in the chamber so that the interior of flasks or vials can be efficiently flushed (Fig. 2.4*b*).

In summary, the previous discussion indicates that three principles should be followed: (1) to avoid short circuits, gases which are more dense than air (e.g., argon) should be introduced at the bottom of a chamber and conversely, less dense gases (e.g., helium) should be introduced at the top; (2) interdiffusion of the incoming gas with that in the chamber is a significant factor, so a fast flush is more effective than a slow flush; (3) experimental results indicate that a chamber is purged at a rate slower than predicted by perfect displacement and faster than that indicated by perfect mixing. Therefore, the perfect mixing equation

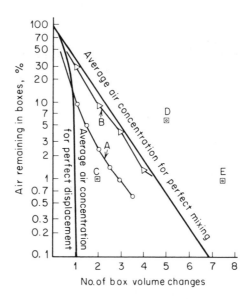

Fig. 2.3. Comparison of experimental results with idealized purge conditions. A, B, and C are experimental, with various designs for the gas inlet and outlet. The best of these, C, was obtained with the purging arrangement illustrated in Figure 2.4*a*. Point D was obtained with the heavy gas (CO₂) entering at the top of the box, and point E was obtained with a light gas (He) entering at the bottom. The latter conditions result in some short-circuit flow. (Reproduced by permission of the copyright owner, Butterworth from P. A. F. White and S. E. Smith, *Inert Atmospheres*, Butterworth, London, 1962, p. 148.)

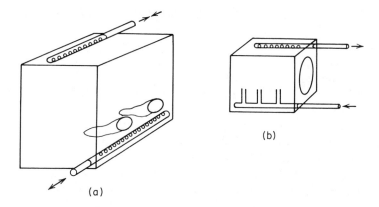

Fig. 2.4. Arrangement of purge lines. (*a*) For the main box the lower manifold should be the inlet with a dense gas (for instance, Ar) and the outlet with a light gas (for instance, He). (*b*) A purge-gas manifold in the transfer port equipped with outlet jets to flush volumetric equipment.

provides an ample margin of safety in estimating the volume of purge gas required. These recommendations are based on the assumption that the chamber can be flushed by smooth lines of gas flow between the gas inlet and outlet. The placement of the gas inlet and outlet to achieve such flow is illustrated in Fig. 2.4*a*.

2.3 SOURCES OF AND REDUCTION OF IMPURITIES IN GLOVE BOX ATMOSPHERES

A. Contamination by Outgassing.   After the initial removal of air from a glove box chamber, the impurity levels will gradually rise due to a variety of sources: (1) outgassing of hydrous or porous materials such as wood, paper, cloth, asbestos, and corroded metal; (2) leakage through seams and joints; and (3) diffusion of atmospheric gases through permeable materials, such as rubber gloves, rubber gaskets, and plastic windows. Items 2 and 3 will be discussed in following section. Because of the first factor, wood or composition board is not suitable for the construction of an inert-atmosphere glove box; and wood, paper, cloth, and asbestos items are not taken into the box when moisture is objectionable. A good substitute for paper or cloth wipes in cleaning up spilled solids is the soft polyethylene sealing film sold by laboratory supply houses (e.g., Parafilm). When paper or cloth wipes must be used, a considerable quantity of moisture may be removed prior to use by storage of these items in a desiccator over a drying agent.

B. Leaks.   Small leaks through seams and gaskets are a significant problem with poorly constructed glove boxes. The maintenance of a small positive pressure in the box does little to reduce the influx of air through small leaks because countercurrent diffusion of air is rapid. To reduce these problems, it has become fairly standard practice to adopt construction practices which approach those used in high-vacuum apparatus. Thus, permanent metal seams are best welded in a smooth and continuous fashion, and permanent seams in poly(methyl methacrylate) are best bonded with a solvent cement (Appendix III). Perhaps somewhat less desirable but still satisfactory seams can be made with rubber cements and rubberized caulking compounds on reasonably matched mating surfaces. This practice suffers from the potential deterioration of the sealing compound. Room temperature vulcanizing silicone sealants are good for most dry box sealing applications. A pitfall to be avoided is to purchase a poorly made glove box or build one with sloppy construction techniques, with the thought that the assembled box can be patched by the application of paints or sealants. In practice, sealants applied over a leak are likely to crack or channel, so one is forever chasing leaks.

Gross leaks may be detected by slightly pressurizing the box and painting suspected areas with soap solution. **NOTE: A normal nonevacuable glove box will tolerate only a small pressure or vacuum before the windows crack, the gloves pop off, or the box is distorted**. The most satisfactory means of finding small leaks is a halogen or helium leak tester (Chapter 7). In this case the box is filled with Freon or helium and the "sniffer" of the leak tester is passed over the suspected areas. Some glove box manufacturers will guarantee a low leak rate as certified by halogen or helium leak testing at the factory.

C. Glove Box Doors and Removable Panels.

Temporary seals cannot be avoided on the entrance doors and on panels which must be removed to reconfigure the glove box. These seals are best made with high-quality firm rubber gaskets or O-rings. It is important that the surfaces which are being joined are sufficiently heavy so that they do not buckle when clamped together, and the panel or door should be clamped evenly. The incorporation of these features in a suitable entrance port is accomplished by the use of a floating hinge that does not constrain the door, and the use of a latching system that exerts its force in an even fashion. Successful door latches involve an autoclave-type closure (Fig. 2.5a), a centrally located screw (Fig. 2.5b), and evenly distributed clamps or bolts (Fig. 2.5c). Large panels are generally bolted together at closely and evenly spaced intervals around the perimeter.

D. Diffusion of Air into the Box.

With good design and workmanship, leakage through seams can be reduced to negligible values. This leaves one remaining major source of atmospheric impurities: diffusion of air through gloves and, to a lesser extent, through rubber gaskets and plastic windows. Diffusion also is a significant source of impurities in plastic glove boxes or glove bags. A few calculations will illustrate the magnitude of the problem.

When the glove box is constructed entirely from plastic, the diffusion of oxy-

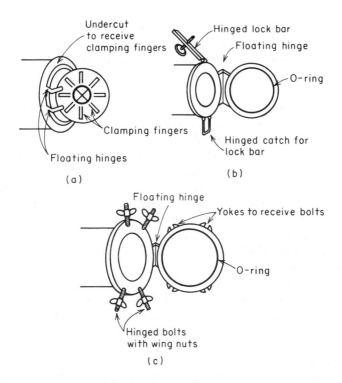

Fig. 2.5. Some designs for antichamber doors. (*a*) Autoclave type. When the hand wheel is turned, the wedge-shaped clamping fingers extend or retract. The door is clamped shut when these fingers are wedged into an undercut on the flange. (*b*) A central screw lock design. (*c*) A hinged-bolt design. There are many variants of this. A group of cam-action clamps is sometimes used in place of bolts.

gen through the walls is significant. For an average-size glove box constructed from 1/2-inch-thick poly(methyl methacrylate), a calculation of diffusion rates indicates that the influx of oxygen through the walls is about equal to oxygen diffusion through the gloves.

Using the diffusion coefficient data in Appendix III, and assuming a typical glove bag size of 8,600 cm^2 and a thickness of 5.2×10^{-3} cm, 12 cm^3 of oxygen will diffuse into the bag at 25°C in one hour and, at a relative humidity of 50%, approximately 30 cm^3 (gas at STP) of moisture will diffuse into the bag in the same time period. For a neoprene glove with an area of 1,900 cm^2 and with a thickness of 7.5×10^{-2} cm, an influx of 0.56 cm^3 of oxygen per hour is calculated. An experimental value for the moisture diffusion rate of 12.5 cm^3 (gas at STP) is available in the literature for 25°C and 50% relative humidity.[4] When the glove box is in use, the worker generates both sweat and heat, and the rate of diffusion of moisture increases by approximately a factor of ten.[4] In other words,

[4] J. E. Ayers, R. M. Mayfield, and D. R. Schmitt, *Nucl. Sci. Eng.*, 8, 274 (1960).

when a dry box with two gloves is idle, the influx of impurities will be about a millimole of atmospheric impurities per hour. Assuming a typical glove box volume of 212 L, this amounts to an increase of atmospheric impurities by 59 ppm/ h. When the box is in use, the influx may increase to about 500 ppm/h.

E. Maintenance of an Inert Atmosphere.

From the previous discussion it is clear that the dry box atmosphere requires continual purification to balance the influx of contaminants. A number of methods of widely varying complexity have been used. The simplest and most common scheme is to maintain a continuous purge of the box. This also may be supplemented with the use of an open dish of desiccant in the box. If the expense of a continuous purge of inert gas is too great, a recirculating system is necessary. The most common recirculators are based on an oxygen scavenger, such as supported Cu or MnO, and a desiccant, such as molecular sieves. (The nature of and regeneration of oxygen scavengers and desiccants are discussed in Chapter 3.) Systems of this type may be constructed, as outlined in Fig. 2.6, or purchased. Commercial systems are available which regenerate the columns automatically. Satisfactory operation is possible with manual regeneration because the unit can be arranged so a set of columns can be switched from purification to regeneration by attending a few valves and switches. One of the simplest commercial units is mounted on the side of the box, and large-diameter columns with large-particle packing are used. An air blower in the box is sufficient to force the inert gas through the column (Fig. 2.7). Mounting the blower in the box simplifies the design because it avoids the leakage which is a problem with many pumps, and it avoids any source of oil mist which can result from a pump. If an external pump is used, a unit with a metal diaphragm is often preferred. Pumps with ordinary shaft seals may be mounted in a metal box which is purged by inert gas.[5,6]

One early test of glove box impurities indicated that a circulation rate of 5 box volumes per hour through the glove box and MnO purification train would lead to a steady state concentration of 50 ppm of atmospheric gases in an argon-filled box. For a modern commercial purification unit the manufacturer claims that the moisture and oxygen levels are reduced below 1 ppm by purging in a large glove box at a circulation rate of about 50 box volumes per hour.

A scheme for removal of nitrogen from a helium-filled glove box can be based on low-temperature adsorption on charcoal or molecular sieves held at $-196°C$. There appears to be no easily regenerated nitrogen scavenger when the more condensable inert gas argon is employed. In this case a getter of hot calcium or hot zirconium-titanium alloy is used. When the column of getter material is consumed, it must be replaced. The heat introduced from this or other sources is an annoyance for the glove box user. This problem may be reduced by including a heat exchanger in the inert-gas stream. Both water-cooled and refrigerated heat exchangers have been used successfully.

[5]E. C. Ashby and R. D. Schwartz, *J. Chem. Educ.*, 51, 65 (1974).

[6]P. A. F. White and S. E. Smith (General References, p. 67).

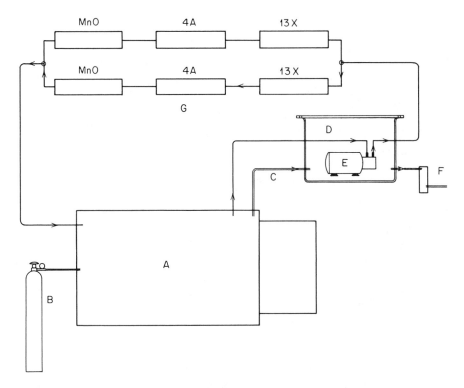

Fig. 2.6. Schematic representation of a dry-box purification scheme. (A) Glove box; (B) tank argon; (C) purge line for pump container; (D) gastight pump container; (E), 2.7 ft³/min graphite ring pump; (F) bubbler; (G) purification train consisting of Linde 13X and 4A Molecular Sieves, and Vermiculite-supported MnO at room temperature (see Chapter 3). In some installations an additional drying column follows the MnO column. Approximate column dimensions are 3-in. diameter by 4-ft length. (Unpublished design of T. L. Brown.)

Hydrocarbons and polar solvents, which are sometimes inadvertently or intentionally introduced into a glove box, present a contamination problem and may have a deleterious influence on some oxygen scavengers. Volatile sulfides and halocarbons are particularly strong poisons for copper-based oxygen scavengers. Removal of these types of materials can be effected by 13X molecular sieves; these in turn can be regenerated at slightly elevated temperatures as described in Chapter 3. Alternatively, the solvent vapors and much of the moisture can be removed by means of low-temperature traps held at $-78°C$.[5]

The pressure in a dry box must be close to 1 atm for easy manipulation of the gloves. Significant deviations from 1 atm for a nonevacuable glove box can damage the windows, gloves, or box walls. If the dry box atmosphere is maintained by a continuous purge, a large mineral oil bubbler on the exit will ensure the proper pressure. With a closed recirculating purifier, a pressure controller is desirable because large fluctuations in the dry box pressure can result from run-

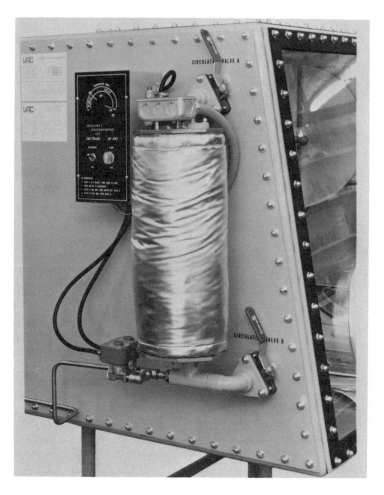

Fig. 2.7. Recirculation unit installed on a dry box. (Reproduced by permission from Vacuum Atmospheres Corp.)

ning the gloves in and out during routine manipulations. Commercial units are available and a description of the necessary parts for the construction of such a unit are given in the literature.[5-7]

F. Monitoring Impurities.   Because of cost and complexity, the automatic monitoring of the glove box atmosphere is not common in academic laboratories, but some industrial and many atomic energy installations employ continuous monitoring. Water is easier to measure than oxygen and fairly simple mois-

[7]D. Eubanks and F. J. Abbott, *Anal. Chem.*, 41, 1709 (1969).

ture meters are available which operate in the parts per million range.[8] These meters are based on a variety of principles, the most common being the conductivity of a hygroscopic material. If the dry box atmosphere will be contaminated by volatile solvents, a meter should be chosen which is insensitive to those solvents. In addition, a simple dewpoint meter may be built into the box in the form of a well which can be charged with a coolant such as Dry Ice.[9] This well should extend down far enough into the box so that its polished surface can be observed for signs of condensation. A dew point of $-78°C$ corresponds to about 0.7 ppm of water vapor.

Oxygen meters are commercially available, but their operation in conjunction with glove boxes presents some complications.[10] One method for the measurement of low oxygen concentrations is based on the potential across a solid oxide electrolyte composed of CaO-doped ZrO_2. Unfortunately, this system works at elevated temperatures where reducing vapors affect the reading. Another scheme involves an electrochemical cell having an aqueous electrolyte.[11] Since this scheme introduces moisture into the gas stream, it is best placed on the outlet from the glove box. Owing to the lack of simple and inexpensive detectors, a variety of qualitative tests are employed. The most popular of these is the light bulb test. In this test, the total oxygen and moisture is estimated by the length of time it takes a light bulb filament to burn out while it is exposed to the glove box atmosphere. The filament of a standard 115-V, 25-W light bulb with a hole broken in the envelope will burn for days to weeks when the oxygen plus moisture contamination is 5 ppm or lower. If the flow rate and nature of the filament are controlled in a reproducible manner, a calibration curve can be obtained; one such curve is presented in Fig. 2.8.[7]

Often it is sufficient to determine oxygen and moisture by noting the changes which occur upon exposure of a highly sensitive compound. **Note: The common cobalt-impregnated desiccants, which turn from blue to nearly colorless when saturated with water, undergo this color change at far too high a moisture level to be useful for sensitive materials.** A qualitative test sometimes recommended for moisture is the brief opening of a bottle of $TiCl_4$ in the glove box. If smoke does not appear, the moisture content is below 10 ppm. Triethyl aluminum or a hydrocarbon solution of diethylzinc will not fume when opened briefly in a box with oxygen in the low parts per million range. These particular tests have the

[8]Sources of moisture meters include Vacuum Atmospheres Co. (see footnote 2); Dupont Instruments, Wilmington, DE 19898; Panametrics, Inc., 221 Crescent St., Waltham, MA 02254; Beckman Instruments, Inc., 2500 Harbor Boulevard, Fullerton, CA 92634.

[9]S. Y. Tyree, Jr., *J. Chem. Educ.*, 31, 603 (1954).

[10]Sources of oxygen meters include: Vacuum Atmospheres Co. (see n.2); Anacon Division of High Voltage Engineering Corp., F.C. Box 416, South Belford St., Burlington, MA 01803; Mine Safety Appliance Co., 600 Penn Center Boulevard, Pittsburgh, PA 15235 (model 803 can be modified for use in the ppm range).

[11]W. Bahmet and P. A. Hersch, *Anal. Chem.*, 43, 803 (1971); D. R. Kendall, *Anal. Chem.*, 43, 944 (1971).

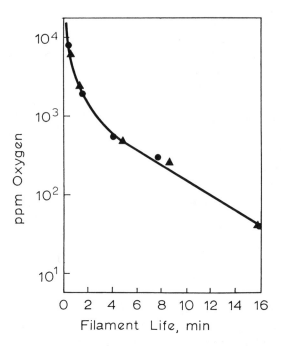

Fig. 2.8. Oxygen concentration versus filament lifetime for an exposed light bulb filament burning in an inert atmosphere. By noting how long the light bulb stays lit, a reasonable estimate of the oxygen plus moisture impurity level can be obtained. These data correspond to an A/C No. 63 bulb filament run with 10 V potential and a gas flow rate of 1.1 L/min.

great disadvantages of contaminating the glove box atmosphere and contributing to undesirable deposits on the glove box walls. If ethers or hydrocarbons can be tolerated in the glove box atmosphere, a solution of $Cp_2TiCl_2ZnCl_2 \cdot$ solvent can be used as a test for oxygen.[12] A small amount of solution is introduced by syringe into an open aluminum pan in the dry box; if the compound retains its original green color as the solvent evaporates, the oxygen content of the dry box atmosphere is less than 5 ppm. At higher oxygen levels the color changes to olive green or yellow; and if the oxygen contamination is very large, it will turn orange. The compound may be prepared by allowing Cp_2TiCl_2 (Alpha Inorganics) to react with a slight molar excess of zinc metal powder in dried and deoxygenated tetrahydrofuran, benzene, or toluene. The reaction may be run by stirring the mixture for approximately 12 hours in a 50-mL Erlenmeyer fitted with a serum cap.

Manganous oxide supported on silica gel has a very high affinity for oxygen, and the reaction is accompanied by a vivid color change from light green to dark brown. A tube of this material in an inert-gas stream is used in the authors'

[12]D. G. Sekutowski and G. D. Stucky, *J. Chem. Educ.*, 53, 110 (1976).

laboratory to measure the oxygen impurity level, and recently a commercial detector of this type has come available for use in gas chromatographic systems. Since the kinetics of this reaction are fast and the line of demarcation is distinct, a measurement of the length of progress of the brown region along the tube can be converted to the amount of oxygen absorbed. The level of impurities can be measured using this scheme by knowing the quantity of manganese on the silica gel, the volume of gas passed through the tube, and the stoichiometry of the reaction, Eq.(4). Although it has not yet been done to the authors' knowledge, a small unit equipped with a simple hand-operated pump could be readily constructed for the periodic check of the atmosphere inside the glove box.

$$2MnO + 1/2O_2 = Mn_2O_3 \qquad (4)$$

2.4 GLOVE BOX HARDWARE AND PROCEDURES

A. Gloves and Glove Ports. Judging from the permeability data in Appendix III, butyl rubber gloves should reduce the influx of water by one-half to one-tenth and that of oxygen by one-third of that given for neoprene gloves. Owing to this advantage, butyl rubber is routinely used on most inert-atmosphere glove boxes. Natural-rubber gloves are sometimes used because they afford a better sense of touch and better dexterity than the synthetics; however, natural rubber is much more permeable to atmospheric gases and less resistant to solvents. To improve dexterity without a large increase in the diffusion of atmospheric gases, gloves are often constructed with a thin cross section in the region of the fingers and thicker rubber elsewhere because the thicker materials have reduced permeability. When dexterity is not critical, some workers use a pair of gauntlet gloves inside the regular gloves. This practice undoubtedly reduces diffusion of moisture into the glove box, and it provides an extra protective barrier when hazardous materials are handled.

The accumulation of sweat in the gloves becomes a great annoyance, which can be alleviated by a small tube carrying air taped into each glove and terminating at the back of the hand, or divided and terminated at the back of each finger opening. Electrician's tape may be used to hold the tube in place. Ideally, the glove should ride in and out with the experimentalist's hands. To improve the adherence of the gloves to the hands, gloves may be mounted palm side up, which results in a slight twist of the glove around the wrist when the hands are in an ordinary working position. Also, to avoid "losing" the gloves, they should be no longer than the worker's reach. This is particularly important when the box is operated at slightly reduced pressures. Some manufacturers offer gloves with accordion sleeves which are supposed to provide free in-out travel of the gloves. In practice the authors find these to be cumbersome, and the complex construction is likely to invite pinhole leaks.

When a new glove box is being designed or an old one modified, the glove ports should be positioned to provide access to the entire bottom of the box and to any shelves or mounting bars on the back wall. With a large box, this may require the use of more than two glove ports. One fairly common practice is to provide a third glove port close to the antechamber side of the box. A glove in this position provides easy handling of the inside antechamber door and allows one to reach into the antechamber.

The glove must be securely fastened to the port so that it does not unexpectedly pop off of the box. One common procedure is to use a large O-ring which holds the cuff of the glove to a groove in the glove port (Fig. 2.9a), and the glove must be further secured by electrician's tape and/or a steel clamping band. If the glove box is of the evacuable type, it is necessary to pump on both the inside and outside of the glove to prevent the gloves from being sucked into the box. Figure 2.9b illustrates a design for a vacuum-tight glove port cover which permits evacuation of the outer side of the glove. The box may be brought up to atmospheric pressure before the outside of the glove is vented, or the two opera-

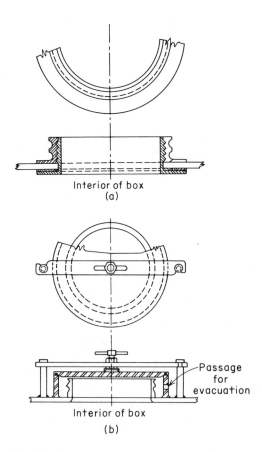

Fig. 2.9. Glove ports. (a) A conventional glove port. (b) Glove port with cover to allow evacuation. (Adapted from USAEC Report TID-16020, N. Garden, Ed., 1962.)

tions may be nearly simultaneous. Once the system is up to 1atm of pressure, the vacuum-tight cover is removed to provide unobstructed use of the gloves.

B. Containment of Hazardous Materials.   To reduce the chance of loss of materials from a containment enclosure, the glove box is operated at slightly below the ambient atmospheric pressure. For glove boxes in which major contamination of the atmosphere cannot be tolerated during the glove changing operation, and for glove boxes which serve as radioactive or biological containment chambers, the glove ports should be designed with a double rib (Fig. 2.9) so that the new glove can be installed before the old one is removed. In preparation for this glove change (Fig. 2.10), the leaky glove is stuffed inside the box and the rib on the cuff of this glove is advanced to the outer groove. A new glove (which may be flushed out with inert gas) is slipped over the old one and its rib is secured on the inner groove. The old glove is pulled off the glove port and into the interior of the box. If this glove is contaminated, it may be bagged out of the box as described below. If radioactive material is involved, the whole operation must be monitored. As mentioned earlier, it is common practice to wear an additional pair of gauntlet gloves inside the dry box gloves to provide additional protection of the experimentalist in the event that a leak develops in the gloves on the inert-atmosphere chamber.

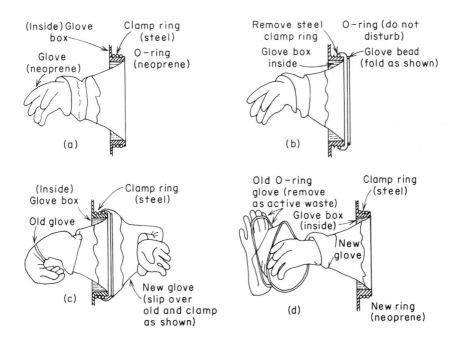

Fig. 2.10.   Changing gloves. The procedure outlined here prevents the loss of radioactivity from the box and minimizes the introduction of air. (Adapted from G. N. Walton Ed., *Glove Boxes and Shielded Cells*, Butterworth, London, 1958.)

The bagging-out process, illustrated in Fig. 2.11, provides a method for discarding waste materials and used gloves which have been contaminated. This process is widely used for radioactive wastes. A heavy polyethylene bag is attached to a double-grooved port similar to a glove port. When the time comes for removal of the wastes deposited into this bag, it is heat sealed and cut along the seal so that the disposable portion of the bag attached to the box is never opened to the atmosphere. A fresh bag is then attached to the inner groove and the remnant of the old bag is taken into the box by the procedure outlined for the glove changing operation, in Fig. 2.10.

2.5 TYPICAL GLOVE BOX SYSTEMS

In this section some glove box systems will be discussed with comments which will put them into perspective for potential applications.

When a laboratory is equipped with good Schlenk-type apparatus and/or vacuum systems, the glove box will be needed for only occasional transfer operations. In this situation a commercial glove bag (Fig. 2.1), may be perfectly adequate for transferring solids. Bulk materials or microcrystalline solids are often much more resistant to oxidation or hydrolysis than solutions. It must be borne in mind, however, that the volume of a glove box or glove bag is large and that prolonged exposure of a sample to a contaminated atmosphere runs the risk of decomposition. Thus it is best to work as rapidly as possible, particularly in a glove bag which is not provided for efficient purging. The mounting of air-sensitive crystals into small glass capillaries for single crystal X-ray diffraction may sometimes be conveniently performed in a glove bag fitted around a binocular microscope (Fig. 2.12). This operation is tedious at best, and the lack of dexterity is particularly troublesome when handling the fragile capillaries. To aid in the manipulation of these capillaries, a capillary holder which is described in Section 2.6.E is a tremendous aid.

The mounting of highly reactive crystals may require the use of a glove box.

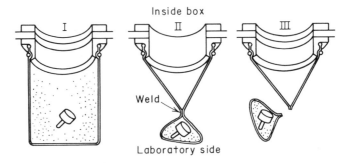

Fig. 2.11. Bagging radioactive wastes. (Adapted from G. N. Walton Ed., *Glove Boxes and Shielded Cells*, Butterworth, London, 1958.)

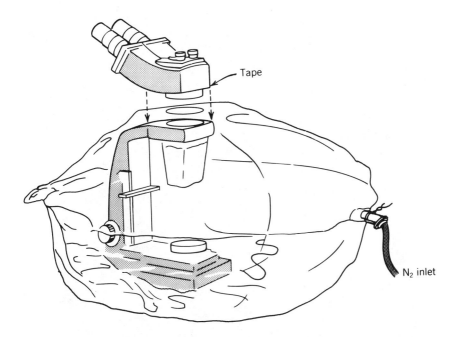

Fig. 2.12. Glove bag fitted around a binocular microscope. A hole large enough to fit the eyepiece of the microscope is cut in the top of the glove bag. The glove bag is then fastened to the microscope using electrical tape and purged of atmospheric gases.

Unfortunately, the working distance for commercial stereo microscopes is rather short, and the microscope also does not accommodate readily the usual large glove box with a vertical or slightly inclined window. A single-chamber evacuable dry box is available commercially which would appear to be satisfactory for crystal mounting and other operations in which the apparatus is not large.[2]

For general use with only moderately air-sensitive compounds, a plastic dry box may be more convenient than a glove bag. Boxes constructed from poly-(methyl methacrylate) or from fiberglass-reinforced polyester are available commercially.[2] As discussed earlier, the diffusion of atmospheric gases through the former material is significant, and it is likely to be less but still appreciable for the fiberglass boxes. These boxes are light and relatively economical and they are useful for all but the most air-sensitive materials.

A commercial welded aluminum dry box, illustrated in Fig. 2.13,[2] has a helium-leak-tested evacuable transfer chamber and an overhead window to provide light. A stainless-steel tray inside of the box protects the aluminum from spills, and a sliding tray in the antechamber facilitates the transfer of materials into and out of this box. The two large doors on the transfer locks are mounted on a pivot arm through a central screw. These doors are equipped with O-ring seals. The O-ring on these doors and similar gasket materials on other glove boxes should be inspected often. This critical seal is subject to the collection of

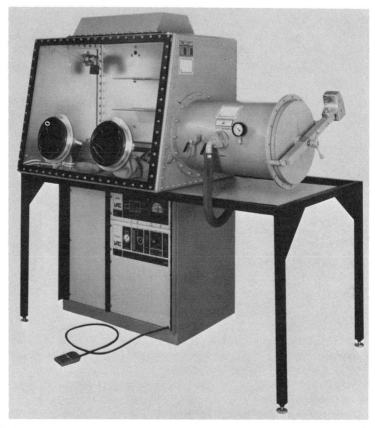

Fig. 2.13. Commercial metal glove box. An aluminum glove box with a recirculating gas-purification system. (Reproduced by permission of the copyright owner Vacuum Atmospheres Corp., North Hollywood, Calif.)

debris from the materials being handled, and it can be nicked if improperly treated. The supplier for this glove box offers efficient recirculating inert-gas purifiers.

2.6 EQUIPMENT AND OPERATIONS

A. Transfer into a Glove Bag. It is cumbersome to transfer large items into a glove bag once work is underway. Therefore all of the items to be used should be placed into the bag, which is then fully deflated and sealed by rolling and clamping the inlet flap. Inert gas may then be introduced and the bag may be flushed through an exit tube and bubbler, or opened slightly, collapsed, and inflated several times to expel air.

B. Example: Transfer of Materials through an Evacuable Air Lock. To introduce a sample from the laboratory the inner door of the port is closed and then the outer door is opened. Samples are placed in the lock and the outer door is clamped shut. The lock is evacuated and filled several times with inert gas. After the lock is brought to atmospheric pressure the final time, the inner door may be opened and samples transferred into the box. Care has to be taken when filling the antechamber to avoid diverting too much gas from the main chamber if the box is not fitted with an automatic pressure control. Another point requiring attention is the nature of the apparatus which is put into the evacuable lock. A closed vessel with standard taper fittings will fly apart under vacuum unless the joints are secured with heavy springs or heavy rubber bands, or unless the apparatus has been evacuated before it is placed in the port. O-ring ware has a big advantage because the clamps used to secure those joints will hold securely under vacuum. Of course, closed containers going into the glove box should have been previously purged or evacuated to eliminate air. As mentioned in an earlier section, wood, paper, cloth, corroded metal, and asbestos are ordinarily not transferred into the glove box because they contain considerable moisture.

C. Weighing Solids. There are several methods for weighing air-sensitive solids. (As mentioned in Chapter 1, this process may be carried out without a dry box if the sample is contained in a sealed ampule of known weight.) The most obvious method of weighing in an inert-atmosphere glove box is to employ a balance which is inside the box. A simple triple-beam balance is very useful for some types of weighing, and a torsion balance or electronic balance is useful when higher sensitivity is required. Static build-up and flexing of the box are two problems encountered in making precise weighings. The antistatic devices mentioned in the following section alleviate this problem to some extent.

Another method for weighing an air-sensitive material is to introduce it into a cappable weighing container in the box and transfer it out to perform the weighing. The weighing tube with cap (Fig. 1.23) is useful when the sample is to be shaken into a reaction vessel on a vacuum line or other apparatus outside the glove box. It is filled and capped inside the box, transferred out and weighed, and quickly opened and shaken into the receiving container under a flush of inert gas. A tare weight is then obtained on the empty weighing tube. Vials filled in the box can be stoppered with a septum cap, removed from the box, and sealed off at the glassblowing bench. Before the seal is made a fine needle is inserted through the serum cap to relieve the buildup of pressure as the ampule is being sealed. To determine the quantity of material in the vial, it is weighed before going into the glove box, the parts are weighed after it is sealed off, and the weight of the contents determined by difference.

D. Static Buildup. The very dry atmosphere of a glove box or glove bag is conducive to the buildup of static electricity, which can scatter a finely divided material, cause small crystallographic-size crystals to fly into oblivion, throw off balance readings, and cause similar vexing problems. There is no complete solution to this problem, but two types of antistatic devices are available which usually reduce the problem to manageable proportions. The head of an alpha-particle static eliminator can be pointed toward the items being manipulated.[13] To provide a sufficient flux of alpha particles, the radioactive source should be replaced frequently. The alpha particles have a rather short range in the glove box atmosphere and they do not pass through the gloves; so the radioactive source has to be placed very close to the work. Another type of static eliminator is the gunlike unit built for removing static from phonograph records. This unit produces ionized gas particles when a trigger is actuated, and these can be "shot" into vessels and the like. Upon long standing inside the dry box these devices lose their effectiveness, so it is common practice to take the static eliminator into the box only when it is needed. It also is worth noting that flint glass is more resistant to static buildup than borosilicate glass.

E. Other Items and Useful Operations. A variety of small items which are generally useful in a glove box are cataloged here. The static eliminators, weighing tube, and Parafilm have already been mentioned. To preserve the cleanliness of the dry box a sheet of aluminum foil is often taken in with each job to cover the working area. Cleanliness is a serious problem, particularly in a communal glove box. If capillaries for X-ray diffraction are kept in the glove box, they should be covered so as to not collect small particles which fly about because of a static charge. A very useful holder for X-ray capillary tubes consists of a glass tube about 15 cm long, with an internal diameter of about 3 mm and a small flare at one end. This flared portion of the tube is lined with a small piece of Parafilm, which does not block the opening in the tube. When an X-ray capillary tube is pushed into the film, it sticks in place and the tube can be grasped firmly without danger to the capillary. Finely drawn glass rods or fine, hard stainless-steel wires are useful for transporting a single crystal into a capillary or for tapping a collection of fine crystals in an X-ray capillary. Other potentially useful items include a collection of spatulas and tweezers. If the box is large, a pair of tongs may help to retrieve items which are out of reach. For infrared spectroscopy on air-sensitive solids, samples may be mulled with mineral oil and sandwiched between infrared transmitting plates. A tissue grinding apparatus is useful as a mulling tool and a supply of degassed mulling oil may be kept on hand in the box for this purpose.

GENERAL REFERENCES

Barton, C. J., 1979, "Glove Box Techniques," in E. S. Perry and A. Weissberger *Techniques of Chemistry*, Vol. 8, Wiley-Interscience, New York, p. 221. A collection of references.

[13]Nuclear Products Co., El Monte, CA.

Barton, C. J., 1963, "Glove Box Techniques," in H. B. Jonassen and A. Weissberger, eds., *Technique of Inorganic Chemistry*, Vol. 3, Wiley-Interscience, New York, p. 259. This chapter stresses the containment of toxic materials and presents a large collection of references on the use of glove boxes.

Sherfey, J. M. *Ind. Eng. Chem.*, 46, 435 (1954); T. R. P. Gibb, Jr., *Anal. Chem.*, 29, (1957); R. E. Johnson, *J. Chem. Educ.*, 34, 80 (1957). Details on the construction and operation of glove box systems.

White, P. A. F., and S. E. Smith, 1962, *Inert Atmospheres,* Butterworth, London. This book presents an excellent discussion of the purification of glove box atmospheres, and a good overview of glove box methods as practiced in the United Kingdom atomic energy establishments. Still the best general reference in the field.

3

INERT GASES AND
THEIR PURIFICATION

Nitrogen, argon, and helium are the most commonly used inert gases for inert-atmosphere work. There are, however, specific situations in which a more reactive gas such as carbon dioxide, hydrogen, or the like is used to exclude air. Of the inert gases, nitrogen has the advantage of being quite cheap and readily available. On the other hand, nitrogen is reactive with some metals, such as lithium, at room temperature. Nitrogen also reacts with many metals at elevated temperatures as well as with some metal complexes. Helium and argon are highly inert and similar in price, so the choice between them generally hinges on the differences in their properties. Helium is the easiest to purify because nitrogen and other gases can be removed by low-temperature adsorption. Argon must be purified by chemical means, since it has low-temperature adsorption properties similar to nitrogen and oxygen. The advantages of argon are that it does not diffuse out of systems as rapidly as helium and it is heavier than air, which makes it easier to maintain an inert atmosphere in cases where a flask must be momentarily opened.

Inert gases are generally available in high-pressure compressed gas cylinders and, in some cases, in medium-pressure Dewars. Details on handling high-pressure compressed gas cylinders are given in Chapter 10. The present chapter will concentrate on the purification of inert gases.

3.1 SOURCES AND PURITY

A. Sources. In the United States high-pressure compressed nitrogen is available in several purity grades. For example, the Matheson Company sells seven grades ranging from "extra dry" (which is quoted to be 99.9% pure) to "prepurified" grade (with 99.998% purity) at relatively economical prices. Other grades, such as "oxygen free" ($O_2 < 5$ ppm) and one called "Matheson Purity" (with the sum of O_2, Ar, CO_2, H_2O, and hydrocarbons equal to less than 5 ppm) are considerably higher in price. The moisture content of the gas will increase as the tank is emptied, because adsorbed moisture exerts a roughly constant partial pressure in the interior of the cylinder. Thus, as the total cylinder pressure is reduced, there will be an increase in the molar ratio of water vapor to gas. High-purity liquid nitrogen is available in most metropolitan areas at prices that work out to be cheaper than that for the compressed gas. For applications which require a large quantity of high-purity nitrogen, such as a continuous glove box purge, large pressurized Dewars of high purity liquid nitrogen are often the most economical source of inert gas. These Dewars are available in gas or liquid take-off designs.

High-purity grades of nitrogen, helium, and argon are often satisfactory for applications in which impurity levels of approximately 50 ppm are tolerable. If relatively large volumes of gas are required and higher purity is needed, it is generally most economical to purify high purity or lower grade gas, as described below. It must be remembered that the purity level is rapidly degraded by diffusion of air through rubber tubing or the rubber diaphragms of ordinary pressure regulators. Special regulators, with metal diaphragms, greatly decrease contamination by atmospheric gases and are available for use on ultra-high purity gas cylinders.

B. Purity Requirements. As an illustration of the purity needed in some specific applications, consider a closed glass apparatus of 500-mL volume. If the inert gas in this apparatus contains 50 ppm of water and oxygen, the total amount of impurity is about 0.001 mmol. This impurity level will be neglegible in a preparative situation if much more than a millimole of compound is being handled. If, on the other hand, the experiment requires a continuous stream of gas, such as in a hot tube reaction or a catalytic flow reactor, 50 ppm of impurities is likely to be unacceptable. It can be seen that it is necessary to estimate the purity requirements on the basis of the amounts of material which are being handled and the total volume of gas to which this material will be exposed. It is not difficult to purify inert gases so that the total oxygen and moisture is below 5 ppm, this being a safe standard for most preparative-scale work. When ultrapure gas is needed (parts per billion range), as in quantitative research on supported catalysts, helium is the inert gas of choice because it is amenable to purification by a combination of low-temperature adsorption as well as chemical methods.

3.2 PURIFICATION OF GASES

A. Principles. A majority of the inert-gas purification schemes employed in the laboratory are based on passing the gas through a bed of solid reactant or adsorbent. In general, the purification process is faster for finer particles and for highly porous particles. Often the adsorption process is controlled by the rate of diffusion through the gas film within the microporous particle, and in this case it is advantageous to employ a long and narrow bed rather than a short bed of large diameter. These conditions of fine particle size and large length-to-diameter ratio increase the pressure drop across the column, so it is necessary to strike a compromise between efficiency and tolerable pressure drop.

Data are available in the literature for gas purification installations ranging from small laboratory experiments to plant-scale operations. To put this information on a common basis, gas velocities are often quoted in terms of the space velocity, which is defined in Eq. (1).

$$\text{space velocity} = \frac{(\text{volume flow of gas at STP})}{(\text{volume of the packing})} \tag{1}$$

The three primary methods for the removal of water from a gas stream are: (1) low-temperature condensation, (2) compression of the gas so that the partial pressure of water increases and condensation results, and (3) the use of drying agents. These methods are sometimes combined. For example, a molecular sieve desiccant will take up more moisture per mole of inert gas at reduced temperatures and at high pressures. The primary method for the removal of oxygen from an inert-gas stream is reaction with a solid or liquid reducing agent. In the case of helium, low-temperature adsorption can be used to remove both oxygen and nitrogen. The following sections present the details of these purification methods.

B. Desiccants. There are significant differences between the water absorption isotherms for substances which form definite solid phases upon interaction with water and for adsorbents and liquid desiccants which undergo a continuous increase in water content without the formation of new phases. For example, $Mg(ClO_4)_2$ forms definite hydrates. As is characteristic with such materials, the absorption isotherm displays plateaus corresponding to the coexistence of two solid phases (Fig. 3.1). An analogous situation is encountered with CaH_2, where the reaction with H_2O produces $Ca(OH)_2$. This leads to a broad region in which the equilibrium vapor pressure of water is constant, and in this case extremely low. By contrast, adsorbents (such as molecular sieves) or liquid desiccants (such as sulfuric acid) display a steady increase in vapor pressure as more water is taken up (Fig. 3.1). The adsorbent-type desiccants also interact with many polar molecules, so competitive binding results with a change in the equilibrium vapor pressure of water and with the total capacity for water. This factor is of little importance when these desiccants are used to clean up an inert-gas stream, but they may be very important when the desiccant is incorporated in a recirculating purifier of a glove box.

Rough equilibrium vapor pressures for H_2O above a variety of common desiccants are given in Table 3.1 along with comments on other properties such as H_2O capacity. Additional factors that are particularly important in the drying of

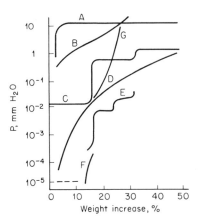

Fig. 3.1. Equilibrium vapor pressures vs. capacity for some common desiccants. (A) Wide-pore silica gel; (B) narrow-pore silica gel; (C) $CaCl_2$, $CaCl_2 \cdot H_2O$, and $CaCl_2 \cdot 2H_2O$; (D) H_2SO_4; (E) Mg-$(ClO_4)_2$ and its hydrates; (F) P_4O_{10} and H_3PO_4; (G) Molecular sieves 4A.

a stream of gas are the flow rate and the physical form of the desiccant. For example, P_4O_{10} is excellent from the standpoint of equilibrium vapor pressure and, as shown in Fig. 3.1, the ultimate capacity is not bad; however, P_4O_{10} is ordinarily supplied as a fine powder which virtually blocks the gas flow if this desiccant is packed into a drying tube. Furthermore, this material is quickly covered with a viscous film of phosphoric acid which drastically reduces the rate of further water uptake. (A supported form of P_4O_{10}, available from the J. T. Baker Co., overcomes some of these problems.) A similar complication is encountered when CaH_2 is employed as a desiccant, because the product of the reaction with water, $Ca(OH)_2$, is produced as a fine powder which may clog a column.

Some desiccants introduce impurities into the gas stream which may be objectionable. This situation is obvious with CaH_2 and other active metal hydrides, where a mole of hydrogen is produced for every mole of water adsorbed. Less obvious is the production of traces of phosphine by the traces of phosphorus(III) impurity present in P_4O_{10}.

CAUTION: Magnesium perchlorate, which from many standpoints is an excellent desiccant, presents a serious explosion hazard in the presence of reducing agents, such as organic materials, CaH_2, etc. This hazard is exacerbated by mineral acids, such as H_3PO_4, or substances which can hydrolyze to acids, such as P_4O_{10}.

Of all the desiccants, molecular sieves 4A or 5A are perhaps the best for all-around use. Two obvious advantages are their high affinity for H_2O and good capacity. They are fairly inert and, because of their size exclusion of large molecules, are less susceptible than many adsorbents to competitive adsorption of other polar molecules. Molecular sieves are usually compacted with a clay binder into the form of small rods, which minimizes the restriction of gas flowing

Table 3.1. Desiccants

Agent	Approximate Water Vapor Pressure at 25°C (torr)	Remarks
CaH$_2$	< 10^{-5}	Evolves hydrogen; no regeneration; basic
P$_4$O$_{10}$	2 × 10^{-5}	Capacity limited by formation of a surface film; acidic; traces of PH$_3$ evolved
Mg(ClO$_4$)$_2$	5 × 10^{-4}	Good capacity; regenerate at 250°C in vacuo; dangerous with reducing agents
BaO	7 × 10^{-4}	Small capacity; regeneration is unhandy; basic
Linde Molecular Sieves, 4A or 5A	1 × 10^{-3}	Good capacity; regenerate at 400°C in vacuo or in a "dry" gas stream
Alumina (active)	1 × 10^{-3}	Fair capacity; regenerate at 500°C in vacuo or in a "dry" gas stream, or 700°C in air
Silica gel (narrow pore)	2 × 10^{-3}	Fair capacity; regenerate at 300°C
KOH	2 × 10^{-3}	Small capacity owing to coating of solid with solution; basic
CaO	3 × 10^{-3}	Limited capacity; especially in the presence of CO$_2$; basic
H$_2$SO$_4$ (concentrated)	3 × 10^{-3}	Oxidizing agent; acidic
H$_3$PO$_4$ (syrupy)	3 × 10^{-3}	Acidic
CaSO$_4$ (Drierite)	5 × 10^{-3}	Regenerated at 250°C
CaCl$_2$	0.2	Good capacity; slightly acidic

through a drying tube. The regeneration of these molecular sieves can be performed in vacuum or inert atmosphere at 400°C.

C. Low-Temperature Trapping and Adsorption of Water.

A low-temperature trap is an effective means of reducing the moisture in a gas to a very low level. Equilibrium vapor pressures given in Table 3.2 demonstrate that at Dry Ice temperature, $-78°C$, the water vapor pressure is approximately 10^{-3} torr. There are, however, difficulties in achieving the equilibrium vapor pressure when a stream of inert gas flows through a low-temperature trap, because fine crystallites of water form in the gas phase and are readily swept through the trap by the stream of gas. Packing the trap with glass helices or glass wool is likely to

Table 3.2. Vapor Pressures (P) of Ice at Various Temperatures (t) [a]

$t°C$	P (torr)
-90	7.0×10^{-5}
-80	4.0×10^{-4}
-70	1.94×10^{-3}
-60	8.08×10^{-3}
-50	2.96×10^{-2}
-40	9.66×10^{-2}
-30	2.86×10^{-1}
-20	7.76×10^{-1}
-10	1.95
0	4.58

[a] E. W. Washburn, *International Critical Tables*, Vol. 3, p. 210.

improve this situation. Also, two cold traps in series afford better trapping because the ice crystallites volatilize between the traps.

Adsorbents also have lower equilibrium water vapor pressures at low temperatures, but this advantage may be offset by a lower rate of moisture uptake. Some isotherms for silica gel and molecular sieves are presented in Fig. 3.1. A trap containing molecular sieves at liquid nitrogen temperature, $-196°C$, is particularly effective in removing moisture from a helium stream, and it also reduces oxygen and nitrogen to low levels. The molecular sieve adsorbent can be readily regenerated by heating in vacuum or in a flowing stream of inert gas as described previously.

D. Adsorption from a Compressed Gas. The efficiency of a desiccant is increased at high pressures. The principle involved is that when a gas is compressed, the partial pressure of moisture in the system is held constant or nearly constant by the desiccant. Thus the molar ratio of moisture to inert gas is decreased. For example, if a desiccant lowers the partial pressure of H_2O to 1 torr, a compressed gas stored over this desiccant at a pressure of 1,000 psig will contain approximately 19 ppm H_2O; at 2,000 psig the gas will contain about 10 ppm H_2O. This technique is used in many commercial air dryers. The process is generally based on an automated two-column system, with the moisture being adsorbed on one column at high pressure while a small amount of this dry gas is diverted to dry out a second column at atmospheric pressure. The gas streams are switched automatically so each column is periodically regenerated. Systems of this type can achieve moisture levels in the parts per million range and the air produced by these dryers is useful for handling moisture-sensitive materials in commercial operations. These air dryers also are commonly used to provide dry air for instruments such as infrared spectrometers and NMR sample spinners.

E. Oxygen Scavengers. A wide variety of reducing agents has been used to remove oxygen from gas streams. When vapors of a solvent can be tolerated, the wet methods have the advantage of being quick and easy to set up. Recipes for the preparation of some common oxygen-removing solutions are listed in Table 3.3. The gas is bubbled through these solutions using a simple gas scrubber or a more sophisticated self-regenerating system (Fig. 3.2). Solutions containing

Table 3.3. Recipes for Oxygen-Removing Solutions

Solution	Preparation
Chromous sulfate (aqueous)	A fresh solution 0.4 M in chrome alum and 0.05 M in sulfuric acid is contacted with lightly amalgamated Zn
Alkaline pyrogallol (aqueous)	15 g pyrogallic acid in 100 mL of 50% aqueous KOH
Sodium hyposulfate (aqueous)	48 g $Na_2S_2O_4$, 40 g NaOH, and 12 g anthraquinone sulfonate in 300 mL H_2O
Sodium anthraquinone sulfonate (aqueous)	2% sodium anthraquinone sulfonate in 1.5 M NaOH is contacted with zinc metal
Benzophenone ketyl (oil)	1 g Na dispersed in mineral oil plus 4 g benzophenone in 1L of mineral oil

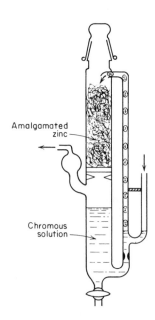

Fig. 3.2. A self-regenerating gas scrubber. The incoming gas causes the solution to percolate through the amalgamated zinc.

the chromous ion have a more favorable rate of oxygen absorption than many other solution scavengers.[1] Significant concentrations of foreign gases, other than the solvent, are introduced by some of these solutions. For example, the chromous ion reacts slowly with water to evolve hydrogen, and alkaline pyrogallol is claimed to evolve CO.

For many applications, the dry oxygen scavengers have the obvious advantage of not introducing solvent vapors into the gas stream. Also, some of the dry scavengers are capable of achieving extremely low oxygen partial pressures. Generally, a column of dry oxygen scavenger is used for long-term applications, such as the purification of the inert gas for permanent Schlenk or glove-box installations. In these cases, the greater inconvenience of setting these up compared to liquid systems is more than offset by the long-term stability and high capacity of the dry columns.

Table 3.4 summarizes some of the most common dry oxygen scavengers. Although a heated tube containing bulk metal, such as copper turnings, is still occasionally used, the bulk-metal oxygen scavengers have been largely displaced by high-surface-area-supported materials. These highly dispersed metal or metal oxide materials react rapidly with residual oxygen and therefore can be used at or near room temperature, which is a tremendous convenience. The procedure for making highly active supported copper metal is available in the literature,[2] and materials of this general type are available commercially.[3] The commercial materials, Ridox and BTS Catalyst, are provided in pellet form, which reduces backpressure in the flowing stream of inert gas. With a space velocity of $6,000 \ h^{-1}$ Ridox should yield purified gas with less than 1 ppm of oxygen at room temperature when the feed stream contains as much as 5000 ppm of oxygen. Literature on the BTS Catalyst indicates a lower but acceptable rate of oxygen removal for particles which are 2–3 mm in diameter. Ideally, the BTS Catalyst should be crushed and screened to the appropriate size, although many research workers do not take the trouble to do this. The Ridox particles are smaller in size and appropriate for use as supplied. One kilogram of BTS Catalyst or Ridox will remove up to 4 L of oxygen at atmospheric pressure and room temperature. Although this capacity can be increased significantly by heating the material to 150°C, this is rarely done. A column at room temperature (which is packed with a 700 × 50-mm bed of the scavenger) will suffice for purifying a dozen or more large high-pressure cylinders of gas.

Supported-copper oxygen scavengers are somewhat self-indicating. The color of the oxidized material is dull green, whereas the active reduced form is dark brown to black. Fig. 3.3 illustrates that the rate of oxygen scavenging decreases as these materials take up oxygen. If oxygen levels on the order of 1 ppm are required, the oxygen removal efficiency should not be allowed to drop below

[1]H. W. Stone, *J. Am. Chem. Soc.*, 58, 2591 (1936).

[2]F. R. Meyer and G. Ronge, *Angew. Chem.*, 52, 637 (1939).

[3]Ridox is available from Fisher Scientific Co., 711 Forbes Ave., Pittsburgh, PA 15219; BTS Catalyst is marketed by Fluka Chemical Corp., 255 Oser Ave., Haupage, NY, 11788.

Table 3.4. Miscellaneous Dry Oxygen Scavengers

Agent	Description
Na-K (67–81% K by weight)	Liquid above 0°C and may be used in a U-tube gas bubbler. Removes O_2, H_2O, and Hg vapor.[a]
Na or K supported on glass wool	Similar to above. Supported Na is prepared by embedding the metal chunks in glass wool, evacuating, and heating to 300°C.[b]
CoO	Removes O_2 at room temperature. Prepared by slowly heating $CoCO_3$ to 340°C in vacuo. Not easily regenerated.[c]
MnO on silica gel	Removes O_2 at room temperature. Easily regenerated. See text for preparation.
Cr^{2+} in silica gel	Prepared by adsorption of a Cr^{+3} solution on silica gel followed by reduction at 500°C in H_2. Efficient, low-capacity O_2 absorption at room temperature.[d]
Palladized or platinized asbestos (or "Deoxo" unit)	Removes traces of O_2 from H_2 at room temperature, removes O_2 from inert gases at room temperature.
Supported Cu (BTS catalyst or Ridox)	70°C required for catalytic removal of O_2 from an H_2 stream, 30–40°C required for catalytic removal of O_2 from CO stream.
Ba, Ca, Ca-10% Mg alloy, La Mg, Th, or Zr	Removal of O_2 from Ar stream at 300, 650, 475, 500, 600, 400, and 600°C respectively.[e] Also removal of O_2 and N_2 at 400, 650, 500, 800, 640, 800, and 1,000°C respectively.[e]
Brass, Cu, Ce, or U	Removal of O_2 from Ar or N_2 stream at 500, 600, 300, 200°C respectively.[e]
Li	Similar to Ca. Removes N_2 and O_2 but reacts with quartz or glass.[f]

[a] E. R. Harrison, *J. Sci. Instr.*, 29, 295 (1952).
[b] H. H. Storch, *J. Am. Chem. Soc.*, 56, 374 (1934); E. R. Harrison, *J. Sci. Instr.*, 30, 38 (1953), supported K.
[c] H. A. Pagel and E. D. Frank, *J. Am. Chem. Soc.*, 63, 1468 (1941).
[d] R. L. Burwell, Jr., private communication.
[e] D. S. Gibbs, H. J. Svec, and R. E. Harrington, *Ind. Eng. Chem.*, 48, 289 (1956); note that hot Mg reacts with Vycor or fused silica tubing.
[f] P. A. F. White and Smith (General References) pp. 48, 222.

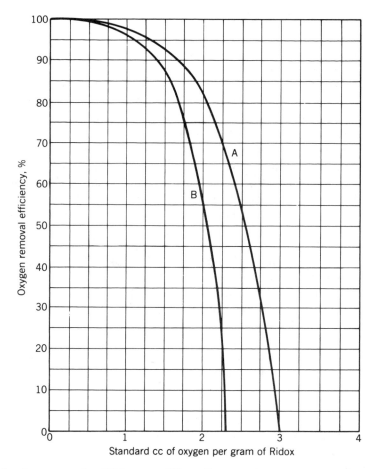

Fig. 3.3. Oxygen capacity of Ridox at two different flow rates. A, space velocity 3000 h^{-1}; B, space velocity 6000 h^{-1}. (Adapted with permission of the copyright owner from Fisher Scientific Co., Bulletin 199B/8-021-01.)

95%. At this level of oxygen impurity in the exit stream, the capacity of the Ridox scavenger would be 1.25 cm^3 of oxygen per gram of scavenger at a space velocity of 3,000 h^{-1}.

Ridox and BTS Catalyst are supplied in the oxidized form and must be reduced at elevated temperatures with hydrogen before use. This process must be done with some care to avoid the excessively high temperatures that cause the supported copper particles to sinter, thus decreasing the efficiency of the catalyst. The regeneration step for Ridox is carried out by wrapping the column with heating tape and heating to 200°C in a stream of nitrogen. The bed is then reduced by the introduction of hydrogen gas diluted with the nitrogen stream. The hydrogen content of this gas should be about 5% by volume. (The gas mixture can be formed by monitoring the hydrogen and nitrogen flow rates with paraffin oil bubblers on the inlet lines. The outlets of these bubblers are joined by a T-tube and the combined stream is conducted to the top of the column.) Sintering of the supported copper will occur if the temperature within the column exceeds 250°C during this process or if the hydrogen concentration exceeds 6%. Alternatively, the regeneration may be performed with 5% hydrogen in nitrogen from a high-pressure cylinder of mixed gas. The temperature conditions are the same as those already mentioned. Temperatures around 150°C are recommended for the regeneration of the BTS Catalyst. The effluent gas from the column should be conducted through a paraffin oil bubbler, and from there to an efficient hood having no open flames. Large amounts of water are produced in the process of regenerating these materials. The bulk of the liquid water is conveniently discarded by way of a valve at the bottom of the column, as illustrated in Fig. 3.4. When the reduction is complete, the last traces of moisture are removed by evacuating the heated column or by continuing to purge dry hydrogen or nitrogen through the heated column and gently heating the base of the column with a hair dryer. Care must be taken to avoid exposing the regenerated column to a sudden influx of air, since this will cause severe overheating. These supported oxygen scavengers cannot be regenerated if they are exposed to gas streams containing organic or metalloid halides, halogens, mercury, sulfides, phosphine, or arsine. In addition, ammonia and most amines are temporary poisons in the sense that they reduce or eliminate the oxygen take-up, but the supported oxygen scavenger can still be regenerated and subsequently used in the absence of these materials.

Manganous oxide is another convenient oxygen scavenger which has been used for the regeneration of inert atmospheres of dry boxes for a long time.[4,5] Recently, it has been shown that a silica gel-supported form of MnO is capable of removing oxygen down to the parts per billion range, and that this material has little tendency to polymerize olefins.[6] These properties, as well as the ease of regeneration, make MnO very attractive for the ultrapurification of inert gases, the purification of olefins to be used in oxygen-sensitive catalytic reactions, and more routine applications. There is evidence that the MnO is poisoned by metal carbonyls at room temperature.

A procedure for the preparation of finely divided MnO is to alternate layers of vermiculite and manganous oxalate in a column similar to that shown in Fig. 3.4. Initial activation is performed by heating the evacuated column to 330°C for 6 h. The resulting green MnO is pyrophoric and must not be exposed to a sudden inrush of air. The predominant reaction with oxygen, Eq. (2), results in a brown-black product; so visual inspection of the column readily indicates its degree of

[4]T. L. Brown, D. W. Dickerhoof, D. A. Bafus, and G. L. Morgan, *Rev. Sci. Instr.*, 33, 491 (1962).

[5]P. A. F. White and S. E. Smith (general references), p. 199.

[6]B. Horvath, R. Moseller, E. G. Horvath, and H. L. Krauss, *Z. anorg. allg. Chem.*, 418, 1 (1975); R. Moeseller, B. Horvath, D. Lindenau, E. G. Horvath, and H. L. Krauss, *Z. Naturforsch.*, 31b, 892 (1976).

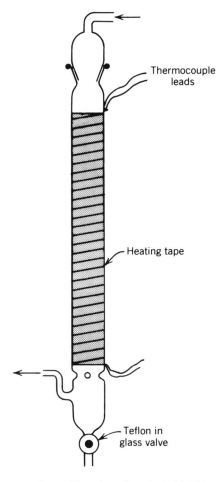

Fig. 3.4. Oxygen scavenger column. The column is packed with Ridox or BTS Catalyst. The heating tape and thermocouple wire are used during regeneration. Water which collects during regeneration can be bled out through the valve at the bottom.

exhaustion. Since fine particles of manganese oxide slough off of this material, a large plug of glass wool is recommended in the exit of the column.

$$6MnO + O_2 = 2Mn_3O_4 \tag{2}$$

Manganous oxide supported on silica gel also is a very clean and efficient oxygen scavenger.[6] The following recipe was designed for small quantities of fine-mesh material to be used in the ultrapurification of helium; it may be scaled up and larger-mesh silica gel may be used for packing large columns. A 105-g quantity of 60–70 mesh silica gel is washed with 1.5% nitric acid, followed by distilled water, and dried at 100°C overnight. A 32-g portion of $Mn(NO_3)_2$ is dissolved in 85 mL of water. With constant stirring, 3-mL portions of this solution are added slowly to the dry silica gel over a period of approximately 1 h. At the end of this process the silica gel should still appear to be dry. Water removal and partial oxidation of this material are performed by heating it overnight at 100°C to produce a black material, which is further oxidized at 300°C in oxygen for 1 h. **CAUTION: Nitrogen oxides are evolved during both of these steps, and these should be vented in a good hood after absorption in base.** If the oxidation was not done in a glass tube, the black solid is packed into a Pyrex tube and then reduced in a hydrogen stream at 370°C for 1 h, yielding a pale-green, extremely air-sensitive solid. The consumption of the MnO by traces of oxygen in an inert-gas stream is evident by a sharp line of demarcation between the green, reduced form and brown-black, oxidized material. When regeneration is necessary, the last-mentioned reduction step is performed. As with the regeneration of supported copper, water is evolved in the regeneration step. This must be eliminated before the column is connected in series with a desiccant.

F.   Nitrogen Scavengers.   The removal of nitrogen from an argon or helium stream is often done with hot titanium sponge. Great care should be taken to remove H_2O from the gas stream before it contacts the titanium. None of these materials can be regenerated, so the charge of nitrogen getter has to be replaced periodically. Commercial nitrogen getter units are available from Vacuum Atmospheres Corp. With a helium stream it is possible to adsorb nitrogen onto molecular sieves at liquid nitrogen temperature, $-196°C$, and thereby achieve low levels of nitrogen. The molecular sieves can be reactivated periodically by warming to room temperature and, if water has been coabsorbed, heating to 400°C to remove this material.

3.3   INERT-GAS PURIFICATION SYSTEMS

Desiccants, oxygen scavengers, low-temperature adsorption traps, and the like are generally combined to form purification systems. Systems of this type used in conjunction with recirculating purifiers for glove boxes have already been described in Chapter 2. This section will provide additional examples ranging from schemes for the routine purification of inert gases to the ppm range to the ultra-purification of helium to the ppb range.

A.   Purification of a Laboratory Inert-Gas Supply.   A typical arrangement for drying and deoxygenating an inert gas from a high-pressure cylinder is shown in Fig. 3.5. The pressure regulator should ideally be of the two-stage type, having an outlet control range of 0–15 or 0–30 psig to permit precise control of the gas in the 2–4 psig range. The valves following the needle valve in Fig. 3.5 are used to isolate the columns from the atmosphere when a tank is being changed, and also to permit flushing air from the regulator and gas line after a new tank is

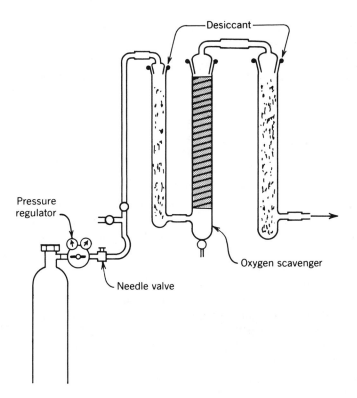

Fig. 3.5. Inert-gas supply with general purpose purification train. Gas contamination is avoided by conducting the gas through metal tubing. Where a flexible connection is desired, a short length of heavy-walled poly(vinyl chloride) tubing is used. Metal bellows valves are included in the inlet line to keep air out while changing cylinders, and to allow air to be purged from the valves and regulator after installing a new cylinder.

attached. This last feature greatly extends the lifetime of the desiccants and oxygen scavenger. Supported copper or MnO is the most convenient oxygen-scavenging material, and molecular sieves 4A or 5A are suitable desiccants. The amount of rubber or poly(vinyl chloride) tubing in the purified gas stream should be kept to an absolute minimum. Where it is needed, heavy-walled poly-(vinyl chloride) tubing or Teflon tubing is preferred because they display low gas permeability and do not degrade rapidly. The columns are generally arranged so that the oxygen scavenging column can be isolated and regenerated without exposing the desiccant to the water which is produced in regeneration. It is possible to regenerate both the oxygen scavenger and molecular sieve columns simultaneously by wrapping them all with heating tape. Alternatively, a single large mixed bed of oxygen scavenger and molecular sieve desiccant can be used, with the regeneration step being carried out long enough at 370°C to eliminate all water.

As noted earlier, these oxygen scavengers are self-indicating, so it is easy to determine when regeneration is necessary. However, there do not appear to be convenient indicators for moisture. The commonly used cobalt-doped desiccants, which change from blue to nearly colorless upon becoming saturated, are insensitive to low levels of moisture and therefore not useful in these columns, except as a precaution against a worst-case situation. The qualitative tests discussed in Chapter 2 are often used for checking a laboratory inert-gas system. For example, the experimentalist may expose a small quantity of aluminum alkyl, titanium tetrachloride, or other sensitive material to the gas as a prelude to performing an experiment. In large installations, continuous monitoring by oxygen and moisture meters may be justified.

B. Ultrapurification of Helium. Although the parts per million range for impurities obtainable in systems described in the preceding section meet or exceed the requirements for the vast majority of laboratory operations, there are situations in which far higher purity is needed. For example, a highly active heterogenous catalyst may have a low concentration of surface-active sites which are readily poisoned by the steady flow of a slightly contaminated inert-gas stream in a flow reactor. In applications requiring very low levels of impurities a sequence of a low-temperature adsorbent, such as molecular sieves or silica gel, plus silica gel-supported MnO will produce helium with reactive impurities in the parts per billion range. One such purification train is illustrated in Fig. 3.6. The low-temperature adsorption traps reduce water and hydrocarbons to very low levels and substantially reduce the level of nitrogen, whereas the MnO trap reduces the oxygen to the low ppb level. The purification train and all subsequent connec-

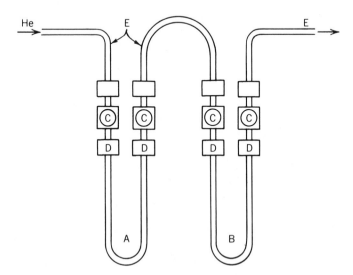

Fig. 3.6. Ultrapurification train for helium. (A) Glass trap containing silica gel which is maintained at $-196°C$ by immersion in a liquid Dewar; (B) glass trap containing MnO-impregnated silica gel maintained at room temperature; (C) Nupro bellows valves, mounted on a sturdy metal plate; (D) Swagelock fittings with Teflon ferrules; (E) copper or stainless-steel tubing. A second low-temperature trap (A) can follow trap (B) to ensure complete H_2O removal.

tions must be very secure to preserve the purity of the gas stream. No rubber or plastic tubing should be used. All valves should be of the metal bellows or diaphragm type (see Chapter 10) and joints either soldered, welded, or of the swage type.

3.4 DETECTION OF IMPURITIES

A series of qualitative and quantitative tests for water and oxygen impurities is given in Section 2.3.F. Of the two, moisture is far easier to detect quantitatively than oxygen.

One of the easiest and most versatile ways to detect oxygen quantitatively in an inert-gas source is to use MnO-impregnated silica gel (described in Section 3.2.E) in which the manganese loading is known. A column of the MnO/SiO_2 material is packed into one arm of a U-tube of known inner diameter and activated. After activation, a known flow rate of inert gas is passed through the MnO/SiO_2 column. The distance the line of demarcation between the black, oxidized material and the pale-green, reduced material travels in a given time is then measured. Using Eq. (3), the amount of oxygen in the inert-gas stream can be determined:

$$O_2(\text{in ppm}) = \frac{580d^2lp(\text{wt \% Mn})}{Ft} \qquad (3)$$

where d is the inner diameter of the glass tubing (in mm), l is the distance the line of demarcation traveled (in mm), p is the density of the MnO/SiO_2 (in g/cc), F is the inert-gas flow rate (cc/time), and t is the elapsed time. As an example, if the MnO/SiO_2 recipe described in Section 3.2.E were used (resulting in a manganese loading of approximately 8%) in a 6-mm-i.d. tube with an inert-gas flow rate of 100 cc/min, and assuming the density of the MnO/SiO_2 is 0.2 g/cc, if the line of demarcation travels 5 mm in 1 h the inert gas contains about 28 ppm oxygen by volume. Much lower levels of oxygen, such as in the ppb range, could be accurately determined by lowering the Mn loading, increasing the inert-gas flow rate, and observing the progressive oxidation of the MnO/SiO_2 for a longer period of time.

GENERAL REFERENCES

Cook, G. A., Ed., 1961 *Argon, Helium and Rare Gases*, Interscience, New York. The purification and handling of rare gases is described here.

White, P. A. F., and S. E. Smith, 1962 *Inert Atmospheres*, Butterworth, London. This book presents an excellent discussion of inert-gas purification, with emphasis on the purification of glovebox atmospheres for large plutonium-handling facilities. Since the appearance of this book better methods of supporting MnO have been developed and various other advances discussed in the present chapter have been made. Nevertheless, this book provides useful general information.

PURIFICATION OF SOLVENTS AND REAGENTS

This chapter describes the removal of water and atmospheric gases from some common solvents and the purification of some common laboratory reagents which are susceptible to contamination by moisture and oxygen. Equipment designs and procedures are introduced here as needed, but most of this material is described in other sections of this book: general inert-gas techniques (1.1, 1.3–1.5), conventional distillation (1.2), desiccants (3.2.B), freeze-pump-thaw degassing (5.3.A), and trap-to-trap distillation (5.3.D,E).

4.1 SOLVENT PURIFICATION

A. General Methods of Purification. Because the solvent often is present in large excess over most reagents, it is important to achieve low levels of reactive impurities in the solvent. Distillation, purging with an inert gas, adsorption, washing, reaction with oxygen or water getters, and freeze-pump-thaw degassing are the most common purification methods.

B. Distillation and Purging. The great efficiency of fractional distillation for the removal of water from hydrocarbon and chlorocarbon solvents is often not well appreciated. The physical origin of this good separation is in part the large positive deviation of the water vapor pressure from Raoult's law because of the lack of affinity of water for these liquids. Typically, the distillation of a simple hydrocarbon solvent with a column of 100 plates and discarding the first two fractions will reduce the water content to levels acceptable for most work. Oxygen removal from solvents which do not react with oxygen is readily achieved by evacuation or purging with an inert gas. The former method is most convenient for materials being handled on a vacuum system and is described in Chapter 5. The purging of solvents with an inert-gas line is readily accomplished using the arrangements illustrated in Fig. 4.1. These methods work well with saturated aliphatic, aromatic, and chlorinated hydrocarbons. Purging methods are not effective with ethers and olefins where peroxide formation occurs through chemical interaction. In these cases, chemical means or adsorbents are needed to purify the solvent prior to distillation.

C. Adsorbents. Water can be removed from solvents by means of adsorbents. Molecular sieves and alumina are the most common adsorbents used to attain very low moisture levels. These adsorbents should be activated prior to use, as described in Chapter 3. The purification can be carried out in a batch manner with the molecular sieves or alumina introduced into a solvent container. After standing and allowing the adsorbent to settle, the solvent may be drawn off. Much lower levels of impurities can be achieved by taking advantage of the multiple-stage purification with a column of adsorbent (see Fig. 1.27). The column is filled with activated adsorbent, then purged with inert gas. The

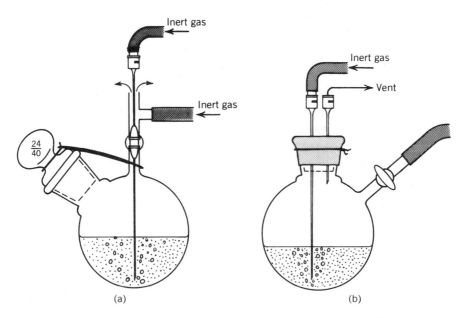

Fig. 4.1. Apparatus for purging solvents with inert gas. (*a*) A long needle attached to an inert-gas source is directed through the sidearm and into the solvent. A slow inert-gas flow runs through the sidearm to avoid back-diffusion of air into the flask. (*b*) Solvent purge using a septum and a double-needle arrangement.

degassed and predried solvent is then admitted to the column, slowly at first. It is best to discard or recycle the first few milliliters of material, and not to exceed the capacity of the column for impurities. A guide to the capacity of active alumina for water is given in the literature.[1] Alumina effectively removes peroxides from alkenes and ethers, but water is adsorbed more strongly than peroxides. Therefore predrying of the solvent in a batch manner with molecular sieves is advisable.

One problem associated with using solid adsorbents to dry solvents is disposal of the used adsorbent, especially if large quantities of solvent are dried. Usually the adsorbent can be reactivated by careful heating in a stream of inert gas or under vacuum. (If the adsorbent was used to dry olefins or other peroxide-containing solvents, any peroxides must first be destroyed; see Section 4.1.D.) Reactivating large quantities of adsorbent, however, can be just as difficult as proper disposal.

D. Prepurification of Solvents. Some solvent purification procedures call for some type of wash. For example, saturated hydrocarbons are often washed with concentrated H_2SO_4 to remove alkenes. These washing procedures are generally done first, since they usually involve the use of an aqueous solution or a substance that contains significant amounts of water. Unlike the following procedures, washes can be done in the presence of air, since further treatment usually involves a step designed to eliminate oxygen.

Bulk water is removed from organic solvents by means of a separatory funnel. Before distillation from a highly active desiccant, very wet solvents should be dried with materials such as $MgSO_4$ or molecular sieve 4A. The solvent is then degassed and charged into the still pot along with the appropriate drying agent. If ethers or olefins are being purified, they should be tested for peroxides before drying. **CAUTION: Peroxides are strong oxidizing agents which may react explosively with strong reducing agents such as $LiAlH_4$ or alkali metals. Peroxides also can be extremely shock sensitive.**

The presence of peroxides in an organic liquid may be detected by the development of an intense brown color when a few drops of aqueous sodium iodide solution are mixed with a few milliliters of the liquid. A more quantitative method involves adding 1 mL of the liquid to an equal volume of a 10% sodium iodide in glacial acetic acid solution. A yellow color indicates a low peroxide concentration, whereas a brown color indicates a high peroxide concentration.

If the concentration of peroxide is not too great, the material can be salvaged by shaking with aqueous ferrous sulfate. In the case of a water-soluble ether, solid CuCl may be used to remove traces of peroxides.[2] Large amounts of peroxides may be removed by storing the liquid over activated alumina or by running the liquid through an activated alumina column. **CAUTION: Do not allow the**

---

[1]M. L. Moskovitz, *Am. Laboratory*, 12, 142 (1980).
[2]*Organic Syntheses*, 45, 57 (1965).

alumina to dry completely. The adsorbed peroxides may be removed by washing the alumina with an aqueous ferrous sulfate solution.

E. Highly Active Desiccants and Oxygen Scavengers.

Although the above methods are sufficiently rigorous for many purposes, there are indications that these procedures still leave trace quantities of oxidizing impurities in the solvent. For example, freeze-pump-thaw techniques do not remove all of the oxygen from benzene as judged by NMR T_1 determinations.[3] For greater purity it is common to employ distillation in conjunction with chemical means for the removal of traces of moisture, oxygen, and oxidizing impurities. A typical solvent distillation setup is illustrated in Fig. 4.2. This particular still is not designed for high distillation efficiency, but rather as a means of separating the solvent from the purifying agents. In this design the reservoir for the distilled solvent is situated above the still pot. The vapor path also permits excess distillate to return to the pot. These features minimize the amount of attention required for the distillation.

It is important to give the still pot an initial charge which substantially exceeds the volume of the upper reservoir so that the still pot will not go dry. It is very important to avoid letting the heated still pot go dry. The heated flask may crack and the solid contents in the pot may decompose to contaminate the system. In some cases an explosion may occur (e.g. when $LiAlH_4$ is used; see below). A metal catch pan should be positioned under each distillation apparatus to contain any spills which may occur upon solvent removal. Each still should also be vented to a fume hood.

Some of the rigorous drying and deoxygenating agents which have been used include $LiAlH_4$, sodium wire or shot, sodium-benzophenone, Na-K alloy, and CaH_2. **CAUTION: These strong reducing agents present an explosion hazard in the presence of chlorocarbons or reducible organic compounds.** Of these agents, solid Na and CaH_2 often have unfavorable rates of reaction with impurities and therefore are not highly effective even at the reflux temperature of the solvent. Although the liquid Na-K alloy and soluble $LiAlH_4$ are often used for the purification of solvents, the authors discourage their use in their laboratories because there is significant risk of explosion if these materials are not handled carefully. Less treacherous dehydrating agents are available. There are many examples of explosions when $LiAlH_4$ has been used to dry ethers, especially tetrahydrofuran.[4] Both peroxides and CO_2 have been discussed as the possible origins of these explosions, and a dry or nearly dry still pot has also been implicated. If $LiAlH_4$ must be used to dry ethers, the peroxides should be removed first, the ether should be predried and degassed, and, in the distillation from $LiAlH_4$, the still pot should never go dry. In addition, a sturdy shield should be used to protect laboratory personnel from the still. Sodium-potassium alloy likewise is some-

[3]J. Homer, A. R. Dudley, and W. R. McWhinnie, *J. Chem. Soc., Chem. Comm.*, 839 (1973).
[4]*Inorganic Syntheses*, 12, 317 (1970).

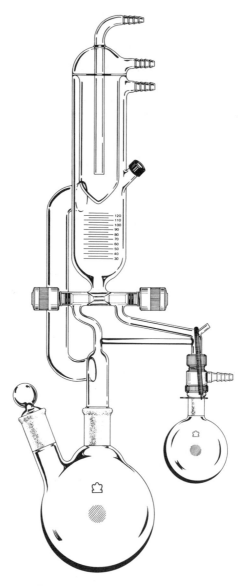

Fig. 4.2   Inert atmosphere solvent still.

what treacherous, because superoxide buildup on the surface of this liquid metal can lead to spontaneous ignition.

Sodium-benzophenone represents a good compromise between ease of use and safety. As mentioned, the solvent should be prepurified to remove most of the water and oxygen, and the still pot should be purged with an inert gas. Sodium wire with a clean surface is then introduced along with a small amount (ca.

5 g/L) of benzophenone. The pot is stirred vigorously with a stout stirring rod to break the residual surface film on the wire. If the solvent is relatively pure, a blue color may begin to form in the solution around the sodium. In the more usual case, it may be necessary to stir for some time to initiate the formation of the blue color. With purging, the pot is then attached to the still, and the solvent is brought to reflux temperature. The solvent should turn a deep blue or green, marking the formation of the sodium-benzophenone and the exhaustion of water and oxidizing species. The distillation is then carried out. If good grades of starting solvents are used, the pot can be replenished with prepurified solvents, with perhaps the addition of some additional benzophenone. The principal drawbacks of sodium-benzophenone ketyl are the extra trouble of initiating its formation and the introduction of traces of benzene into the dried solvent.

4.2 SOLVENT STORAGE

Freshly distilled solvents are readily contaminated and, in the case of ethers which are supplied with peroxide inhibitors, the purification step leaves the solvent susceptible to rapid peroxide formation. Therefore, it is necessary to provide the proper means of solvent storage. It is possible to minimize the amount of fresh solvent which must be stored by using the distillation setup illustrated in Fig. 4.2, which has a reservoir that may be tapped when needed. A simple side-arm flask (see Fig. 1.3) is satisfactory for short-term use when solvent purity is not highly critical. Solvent may be removed from such a flask by syringe through the open joint while a flush is maintained on the equipment, or a septum side-arm may be used. Much better exclusion of the atmosphere can be achieved by means of a Teflon valve closure, such as illustrated in Fig. 4.3. The advantages of this design are the excellent isolation of the solvent by the Teflon-tipped valve and the lack of contamination of the solvent by stopcock greases. This type of solvent storage tube is ideal for use with a vacuum system and is also satisfactory for removal of solvent by syringe while an inert-gas flush is maintained from the sidearm of the valve.

4.3 DETECTING IMPURITIES

Peroxide detection in organic solvents was previously discussed in Section 4.1.D. It is imperative that solvents suspected of peroxide contamination be tested, since an extreme explosion hazard results when peroxides are concentrated in the still pot during distillation.

The detection of water in solvents, especially in trace quantities, can be troublesome. A method employing infrared spectrometry has been developed by Barbetta and Edgell for the detection of water in solvents at the submillimolar

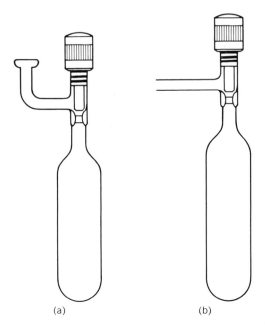

(a)   (b)

Fig. 4.3. Solvent storage tubes with Teflon valves. (*a*) The O-ring joint on the sidearm allows attachment to a vacuum line equipped with O-ring joints. (*b*) Same basic design as in (*a*) with a straight sidearm. This design is easily connected to a Schlenk line via Tygon or other flexible tubing that fits over the sidearm.

level.[5] In this method, the intense water fundamental absorption band at 3,590 cm^{-1} was used to monitor the water content of a variety of solvents, including acetonitrile, THF, and pyridine.

4.4 PURIFICATION OF SPECIFIC SOLVENTS

A. Water. Purging with inert gas or pumping on this solvent will rid it of dissolved oxygen. Dissolved salts may be removed by distillation or by utilizing a commercially available deionizing system.

B. Saturated Hydrocarbons. Purging and freeze-pump-thaw methods are effective in degassing paraffins. Distillation with the elimination of the first fractions produces dry solvent. Molecular sieves may also be used to remove water. Distillation from sodium-benzophenone ketyl is highly effective in obtaining very low levels of water and oxygen. Commercial grades of hydrocarbons are often contaminated with olefins, which can be eliminated by several washings

[5]A. Barbetta and W. Edgell, *Appl. Spect.*, 32, 93 (1978).

with concentrated sulfuric acid, separation from the acid, washing with water, and, after preliminary drying with molecular sieves 4A or 5A, subjecting the solvent to fresh desiccant or one of the drying procedures outlined above.

C. Aromatic Hydrocarbons. The methods listed for saturated hydrocarbons may be used with benzene. Thiophene and similar sulfur-containing impurities are removed by sulfuric acid washes. Very-high-purity benzene may be prepared by fractional crystallization from ethanol followed by distillation.

Toluene and xylenes are purified in the same manner as benzene; however, these solvents should be kept cool (at or below room temperature) during sulfuric acid treatment due to their greater reactivity toward sulfonation.

D. Chlorocarbons. Molecular sieves provide a good means of drying these materials. **CAUTION: Strong reducing agents, such as metal hydrides, Na, and Na-K, react violently with halogenated hydrocarbons and should never be used to dry these materials.** Chloroform generally contains 1% ethanol to suppress phosgene formation. This may be removed by shaking with concentrated H_2SO_4, separating from the acid, washing with water, predrying with molecular sieves or silica gel, drying with molecular sieve 4A, and distilling, taking the central fraction. This material must be used directly or stored strictly out of contact with the atmosphere. Methylene chloride is less reactive than chloroform and is not prone to phosgene formation. It may be distilled from P_4O_{10} or dried with molecular sieves and distilled under an inert atmosphere. (Note that traces of phosphine are introduced into a material by the use of P_4O_{10}.)

E. Ethers. Diethyl ether is available in very dry grades, which for most purposes can be used directly from the freshly opened can. Similarly, purified grades of tetrahydrofuran are available which are often sufficiently dry to be used directly (Fisher). When extremely sensitive materials are handled, diethyl ether and tetrahydrofuran are often distilled from $LiAlH_4$. However, great care must be taken first to remove all peroxides and never to let the still pot go dry (see the general discussion in Section 4.1). Sodium-benzophenone ketyl is also effective with ethers which are peroxide free, and it has the advantage of being safer than $LiAlH_4$. A small amount of benzene is introduced into the solvent when sodium-benzophenone ketyl is used.

F. Nitriles. Acetonitrile is a convenient solvent for many ionic compounds, but it is extremely difficult to dry to much better than millimolar in water. In addition, degradation products of the solvent, ammonia and acetic acid, may be present. Strongly acidic or basic drying agents react with this solvent and strong reducing agents are ruled out. One recommended procedure is first to predry the solvent with molecular sieves or silica gel if large quantities of water are present,[6]

[6]J. F. Coetzee, *Pure and Appl. Chem.*, 13, 429–433 (1966).

then to stir or shake this material with calcium hydride, which will remove both acetic acid and water. The predried acetonitrile is then distilled at a high reflux ratio from P_4O_{10} (less than 5 g/L) under an inert atmosphere. A gel may form in the distillation flask, and care should be taken to leave behind the bulk of the solvent and residues in the distillation flask. The solvent is then refluxed over calcium hydride and distilled under an inert atmosphere using a high reflux ratio and a good fractionating column. The middle fraction is collected. Acrilonitrile is an impurity which will contaminate the product if this distillation is not efficient.

As an alternative to this procedure, some workers advocate stirring acetonitrile with calcium hydride, distilling it under a dry inert atmosphere, and passing it through a highly activated alumina column under a dry inert atmosphere.

G. Alcohols. The removal of traces of water from methanol and ethanol is difficult. Calcium turnings, magnesium turnings activated with iodine, and lump calcium hydride are sometimes used to remove traces of water from methanol. However, these active metals may contain some nitride and give rise to contamination by ammonia. These same drying agents may be used with higher alcohols. A column of dry molecular sieve 4A is also effective in removing water from ethanol and higher alcohols.

H. Acetone. Acetone is very difficult to dry, since many of the usual drying agents cause reaction, including $MgSO_4$. Storage over molecular sieve 4A yields relatively dry acetone. High-purity acetone may be obtained by saturating the solvent with dry NaI at room temperature, decanting and cooling to $-10°C$, isolating the crystals which form (NaI-acetone complex), warming the crystals to room temperature, and distilling the resulting liquid.

4.5 PURIFICATION OF SOME COMMONLY USED GASES

A. Acetylene. This gas is sold in cylinders containing acetone, which maintains a low pressure of acetylene. Therefore, acetone is the principal contaminant and may be removed by passing the gas through an aqueous solution of $NaHSO_3$. The gas stream is then dried with molecular sieves. When small amounts of acetone-free acetylene are needed, it is convenient to generate the gas from calcium carbide and water. Again, the resulting acetylene stream is dried with molecular sieves.

CAUTION: **Acetylene is thermodynamically unstable with respect to the elements and it may explode spontaneously at high pressures. The gas should never be handled above 15 psig pressure. If water addition to calcium carbide is used to generate acetylene, the generator must be kept cool and the acetylene allowed unrestricted flow from the generator.**

B. Ammonia. This good low-temperature or high-pressure solvent may be purchased in a fairly dry state. The last traces of water can be removed by condensing the ammonia onto sodium on a vacuum line, allowing the blue color of dissolved sodium to develop, and trap-to-trap distilling of the resulting ammonia into a metal cylinder or into a reaction vessel. The experimentalist should be thoroughly familiar with vacuum line operations before attempting this procedure. The major pitfall is bumping the ammonia with a sudden pressure rise, which can propel stopcocks at great speeds or burst the glass apparatus. One other potential problem is contamination of the vacuum line with sodium resulting from a fine spray which forms as the ammonia evaporates. A generous plug of glass wool above the drying tube reduces this last problem, and a pressure release bubbler (see Fig. 1.2) will reduce the chances of pressurizing the vacuum system.

C. Boron Halides. These compounds are best handled and purified on a vacuum system. Silicone stopcock grease reacts with the boron halides, and most hydrocarbon greases contain unsaturated hydrocarbons which are polymerized by these Lewis acids. A halogenated stopcock grease, Teflon-in-glass valves, and Viton O-rings all give satisfactory performance with the boron halides. Boron trifluoride is available in cylinders and may be freed from residual atmospheric gases by trap-to-trap distillation in the vacuum system. For very critical applications, further purification can be achieved by the formation of the benzonitrile adduct at $0°C$ in a vacuum system; this adduct is then evacuated to remove volatiles such as the weaker acid SiF_4. The BF_3 is recovered from its benzonitrile adduct by thermal decomposition.[7] Adjacent to the trap containing the adduct is a trap cooled to $-78°C$ (Dry Ice), and this is followed by a trap cooled to $-196°C$ (liquid N_2). About 20 torr of dry nitrogen is introduced into this train of three traps; the first trap containing the adduct is warmed to about $60°C$ with hot water and the valve leading to the adjacent, $-78°$ trap is cracked open slightly, while the valve between the $-78°$ and $-196°$ trap is opened. In this way benzonitrile collects in the $-78°$ trap and BF_3, in the $-196°$ trap. When the decomposition of the adduct is complete, the purified BF_3 is trap-to-trap distilled to a storage bulb.

The contaminants in commercial boron trichloride usually are HCl and $COCl_2$, as well as oxychlorides which have some volatility. The oxyhalides are readily removed by one or two trap-to-trap distillations in a clean vacuum system. Most of the HCl can be removed by holding the BCl_3 at $-78°C$ and pumping away the volatiles for a brief period (some BCl_3 is sacrificed in the process). Phosgene, which may be detected by its gas-phase infrared absorption at 850 cm^{-1}, is very difficult to remove. Liquid boron tribromide is generally supplied in sealed ampules. If it is straw colored, dibromine is a likely impurity, and this

---

[7]H. C. Brown and R. B. Johannsen, *J. Am. Chem. Soc.*, 72, 2934 (1950).

can be removed by shaking the compound with mercury. Hydrogen bromide may be removed by trap-to-trap fractionation in which the BBr_3 is collected at $-45°C$ and the HBr at $-196°C$.

D. Carbon Dioxide. Water is generally the main contaminant in carbon dioxide, especially if Dry Ice is used as the source of this gas. Trap-to-trap sublimation of the carbon dioxide through an intervening activated silica gel trap on a high-vacuum line effectively removes water. Low-boiling gases such as nitrogen or oxygen are removed if the sublimation is done under dynamic vacuum.

E. Carbon Monoxide. The most troublesome impurity in carbon monoxide generally is oxygen, which may be removed by passing the gas over supported copper (e.g., Ridox) or supported MnO, both of which are discussed in Chapter 3. It is reported in the literature that CO does not impair the performance of MnO, but experience in the authors' laboratory indicates that MnO that has been exposed to CO is not readily regenerated.

F. Ethylene. Oxygen is a very troublesome impurity for many studies and can be removed by passing the ethylene over supported MnO or supported copper.

G. Hydrogen. Oxygen may be reduced to the low parts per million range by means of a platinum-metal catalyst, which combines the residual oxygen with hydrogen.[8] The ultrapurification of hydrogen can be achieved by diffusion of the gas through a heated palladium thimble. Commercial units are available that are based on this principle.[8]

H. Hydrogen Halides. Water can be removed from HCl and HBr by trap-to-trap distillation in a vacuum system. If these gases are handled at 100 torr or below, a trap held at $-78°C$ may be used to condense water. An activated-carbon trap can be used to remove any halogen impurities in these gases. Performing the transfer under dynamic vacuum will also remove any low-boiling gases not condensable at liquid nitrogen temperatures.

The purification of HI is complicated by the well-known equilibrium between itself and hydrogen and iodine, Eq. (1).

$$2HI(g) \rightleftharpoons H_2(g) + I_2(g) \tag{1}$$

The attainment of equilibrium is very slow at room temperature. Residual iodine can be removed by passing the gas over red phosphorous or shaking it with mercury. The gas may be then be passed through a trap at $-78°C$ to con-

---

[8]Available from Matheson Gas Products, P.O. Box 85, East Rutherford, NJ 07073.

dense moisture, and condensed in a second trap at $-196°C$. A high vacuum is maintained on the exit of the second trap to remove hydrogen.

I. Sulfur Dioxide. This compound is a useful reactant and low-temperature solvent. The commercial compressed gas may be condensed at liquid nitrogen temperature and atmospheric gases pumped away. Moisture is removed by passing it through a trap containing P_4O_{10} dispersed in glass wool.

4.6 ANHYDROUS METAL HALIDES

These very useful starting materials often are reactive with atmospheric moisture. Simple dehydration by the application of heat is not effective when the metal ion is small and/or in a high oxidation state because hydrogen halide is eliminated and metal oxyhalides or hydroxyhalides result. Several general methods are available for the removal of oxygen-containing ligands from metal halides; but elaborate schemes for the purification of commercial halides are not attractive, because the direct synthesis from pure starting materials is generally less difficult. The cumulative indices of *Inorganic Syntheses* may be consulted for these synthetic procedures.

Commercial grades of aluminum chloride and aluminum bromide are generally contaminated with hydroxylic impurities. Some purification can be achieved by subliming the material in a glass apparatus under vacuum. Better purification can be accomplished by mixing good-quality halide (e.g. that sold by Fluka) with aluminum wire and 1%-by-weight sodium chloride prior to the sublimation. The function of the NaCl is to form molten $NaAlCl_4$, which has an affinity for ionic impurities.[9] A second sublimation is sometimes necessary to remove fine particles of aluminum if the first sublimation was performed too rapidly. The purest samples of aluminum halides are obtained by direct synthesis.

A variety of transition metal chlorides and other chlorides can be dried by means of thionyl chloride, with the evolution of SO_2 and HCl.[10] Twenty grams of the finely divided metal chloride is placed in 50 mL of freshly distilled $SOCl_2$. When the evolution of gas has ceased, the mixture is refluxed for 1–2 hours, and then excess thionyl chloride is distilled from the reaction flask under reduced pressure. Residual thionyl chloride is removed from the product in a vacuum desiccator containing KOH. The halide should be transferred and stored in a moisture-free atmosphere. Halides which have been successfully dried by this method include: LiCl, $CuCl_2$, $ZnCl_2$, $CdCl_2$, $ThCl_4$, $CrCl_3$, $FeCl_3$, $CoCl_2$, and $NiCl_2$.

---

[9]J. Robinson and R. A. Osteryoung, *J. Am. Chem. Soc.*, 101, 323 (1979); see also D. W. Seegmiller, G. W. Rhodes, and L. A. King, *Inorg. Nucl. Chem. Let.*, 6, 885 (1970), for a recrystallization technique.

[10]A. R. Pray, *Inorganic Syntheses*, 5, 153 (1957).

GENERAL REFERENCES

Braker, W., and A. L. Mossman, 1980, *Matheson Gas Data Book*, 6th ed., Matheson Div. Searle Medical Products, Lyndhurst, N.J., 07071. Although this volume does not present methods of purification, it does provide information which is useful in the handling of gases, such as tabulations of physical properties, infrared spectra (sometimes showing impurities which are not identified as such), safety information, and handling instructions.

Dodd, R. E., and P. L. Robinson, 1954, *Experimental Inorganic Chemistry*, Elsevier, Amsterdam. Properties and methods of preparation and purification for many inorganic compounds are given here.

Gordon, A. G. and R. A. Ford, 1972, *The Chemists Companion*, Wiley-Interscience, New York, p. 429-439.

The series entitled *Inorganic Syntheses* generally describes purification methods in conjunction with synthetic procedures. Cumulative indices should be consulted in volumes 10, 15, 20, and the most recent issue. Volumes 1-17 were published by McGraw-Hill, New York; Volumes 18 to present are published by Wiley, New York; back volumes are available from Kreiger Publishing Co., Box 9542, Melbourne, Fla., 32901.

Jolly, W. L., 1970, *Synthesis and Characterization of Inorganic Compounds*, Prentice-Hall, Englewood Cliffs, N.J., p. 114-121.

Lagowski, J. J., Ed., *The Chemistry of Non-Aqueous Solvents*, Academic, New York. This series contains detailed accounts of the purification, properties, and handling of some major solvents: Vol. 2 (1967), hydrogen halides, amides, and ammonia; Vol. 3 (1970), sulfur dioxide and acetic acid; Vol. 4 (1976), tetramethylurea, cyclic carbonates, and sulfolane; Vol. 5A (1978), trifluoroacetic acid, halosulfuric acids, interhalogens, inorganic halides and oxyhalides.

Perrin, D. D., W. L. F. Armarego, and D. R. Perrin, 1980, *Purification of Laboratory Chemicals*, 2nd ed., Pergamon, London. This book is broader but less incisive than the one listed above. General purification techniques are discussed, and very brief directions are given for the purification of a wide variety of organics and inorganics.

Riddick, J. A., and W. B. Bunger, 1970, *Organic Solvents, Techniques of Chemistry*, Vol. 2, 3rd ed., Wiley-Interscience, New York. A very good summary of the various techniques reported for the purification of individual solvents along with some physical data and references to the original literature.

Vacuum Line
Manipulations

5

VACUUM LINE DESIGN
AND OPERATION

When properly handled, a vacuum line is a true joy to use because it provides a closed reaction vessel with excellent exclusion of air, quantitative retention of all reaction products, and a ready means for the transfer and quantitative measurement of gases.

Vacuum lines are employed in problems ranging from the simple transfer of volatile substances to the extended series of operations involved in the synthesis, purification, and characterization of new compounds. This chapter presents an overview of the design of glass vacuum systems. Following chapters give greater detail on individual components such as pumps (Chapter 6), manometers (Chapter 7), joints and valves (Chapter 8), and metal vacuum systems (Chapter 10). The present also describes the basic operations for the transfer and purification of condensable and noncondensable gases, the quantitative PVT (pressure-volume-temperature) measurement of gases, and the characterization of condensable gases by vapor pressure. More specialized operations are described in Chapter 9.

5.1 GENERAL DESIGN

A. Typical Systems. A simple system for the transfer of samples to an infrared gas cell or to a NMR sample tube consists of a fore pump, diffusion pump, trap, and manifold (Fig. 5.1). At the other extreme is a general-purpose chemical vacuum line, which permits the separation of volatile compounds, transfer of noncondensable gases, and storage of reactive gases and solvents (Fig. 5.2). When attack of stopcock grease is a serious problem, grease-free de-

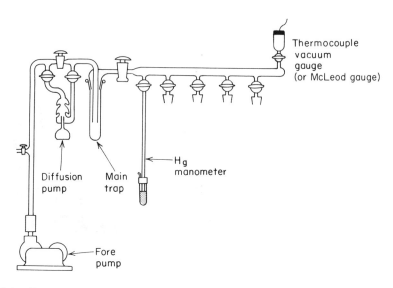

Fig. 5.1. Simple vacuum line. A line of this sort might be used to transfer volatile samples to infrared gas cells, NMR tubes, etc. It might also be used for loading tubes for sealed-tube reactions and for vacuum sublimations.

signs based on O-ring joints and glass valves with Teflon stems are frequently used (Fig. 5.3). Each stopcock, valve, or joint adds to the cost and to potential sites for leakage, so the vacuum line design should be the simplest possible to serve the intended uses.

B. Major Components. The extended array of glass tubing in a typical vacuum system requires proper support to minimize breakage. A rigid lattice is essential to reduce the differential movement of various glass components. For large vacuum systems the lattice is generally firmly mounted on a sturdy low bench as illustrated in Fig. 5.4. Horizontal bars increase the rigidity of the lattice, but they should not be used for clamping apparatus. **NOTE: It is important to clamp the vacuum line parts only to vertical bars, because a clamp may swivel around a horizontal bar and thus not provide reliable support.** Heavy permanent items, such as mercury-filled Toepler pumps or McLeod gauges, are preferably mounted on the bench and supported in a bed of plaster of Paris. When the vacuum line is first assembled and carefully but firmly clamped in place, it is good practice to anneal segments of tubing between adjacent clamps to relieve stress on the apparatus.

The functional units in the general-purpose vacuum line illustrated in Fig. 5.2 are fairly typical: (1) a source of high vacuum, (2) a high vacuum manifold, and (3) various working manifolds. These working manifolds may contain a train of U-traps for the separation of volatiles, or a series of containers for the storage of either gases or volatile liquids. In addition to U-traps, the working

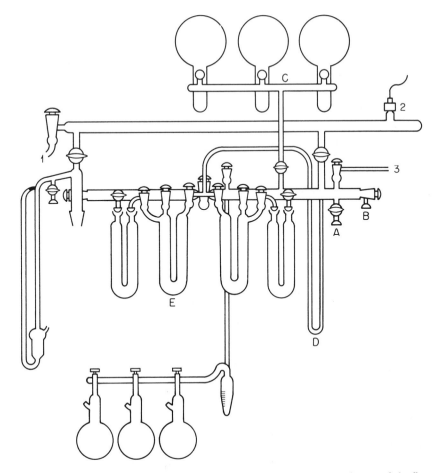

Fig. 5.2. Multipurpose high-vacuum system. Bulbs for gas storage are at the top of the figure. Directly below these is the high-vacuum manifold, which is attached to a trap and pumps at site 1 and connected to a high-vacuum gauge at 2. The working manifold in the center of the figure is attached to a train of U-traps and to various inlets. It is also connected to a Toepler pump at 3. Gases and volatile liquids are introduced through the inlets attached to this manifold, and large reaction vessels or special apparatus are frequently attached to the large joint. At the bottom of the figure is a manifold with solvent storage vessels attached; it is also connected to a graduated trap which allows measurement of the volume of solvent before it is transferred to the main working manifold. Standard high-vacuum stopcocks are employed except on the solvent and gas storage bulbs, where greaseless needle valves are used.

manifold generally has sites for attaching reaction vessels, gas inlets, and the like. This working manifold usually includes one or more mercury-filled manometers for the PVT (pressure-volume-temperature) measurement of gaseous reactants or products, or for vapor pressure measurements. The main high-vacuum manifold is usually equipped with a high-vacuum gauge so the initial degree of evacuation can be measured. This gauge is also valuable for locating

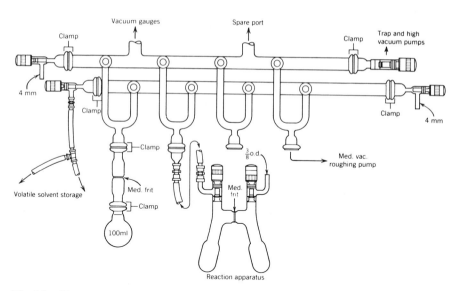

Fig. 5.3. Wayda-Dye greaseless vacuum line. This apparatus makes extensive use of metal bellows tubing and O-ring seals. Thus the reaction vessels, filters, and other items can be tilted and manipulated like Schlenk ware, and high vacuum conditions can be achieved for the removal of atmospheric gases and for baking out residual moisture. Trap-to-trap distillation of volatile solvents such as NH_3 or SO_2 is readily accomplished with this apparatus. This version is not designed for the measurement of volatiles or trap-to-trap separation. (Reproduced from A.L. Wayda and J. L. Dye, *J. Chem. Educ*, *62*, 356 (1985) by permission of the copyright owner the Division of Chemical Education of the American Chemical Society.)

leaks in the system. A Toepler pump is generally attached to the vacuum line if noncondensable gases are to be manipulated. The Toepler pump is not an absolute necessity for transferring noncondensable gases such as CO and CH_4. As described in Section 5.6.B, low-temperature adsorption provides a simple method for transferring these gases.

The mechanical fore pump is a heavy item which sets up considerable vibration. This vibration disrupts the menisci of mercury manometers and is otherwise undesirable. To minimize the transfer of vibrations to the vacuum line, the fore pump generally is mounted on the floor (rather than on the bench of the vacuum rack) and the connection between the fore pump and the vacuum system is made with heavy-walled vacuum tubing or flexible corrugated metal tubing.

Details on pumps, manometers, vacuum gauges, special apparatus, and leak testing are given in Chapters 6–10. It is the purpose of the remainder of this chapter to describe the transfer of condensable and noncondensable gases, trap-to-trap fractional separation of volatiles, and the use of vapor pressure in the characterization of volatile compounds. These operations are basic to practically all chemical vacuum line work.

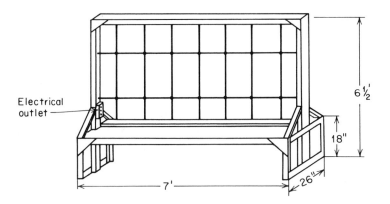

Fig. 5.4. Typical lattice. Glass apparatus is clamped to the vertical rods, which are spaced at 10- to 12-in. intervals. One or two horizontal rods are generally included to impart rigidity. When the rack is large, $1/2$-in.-diameter steel or stainless-steel rods are preferred to aluminum rods because the latter do not afford a rigid lattice. Frames are constructed from pipe (about 2-in. diameter), welded angle iron, welded H-beam, bolted Flexi-frame, or similar rigid members. The vacuum line lattice should be grounded by running a copper wire from the lattice to a water pipe. It is advantageous to cover the bench with Transite or a similar heat-resistant surface, since this facilitates glassblowing. The dimensions given are typical of a rather large rack. A hooded vacuum rack is occasionally used to help contain reactive or poisonous substances released in a mishap. To provide the required rigidity, it is sometimes necessary to brace the top of the vacuum rack to a wall.

5.2 INITIAL EVACUATION

A good initial vacuum is necessary because noncondensable foreign molecules will impede the movement of the condensable material. When large quantities of relatively unreactive gases are being handled (for example, BF_3), it is usually satisfactory initially to evacuate the system to 10^{-3} torr. Small quantities of gas and highly reactive gases require a better initial vacuum for their transfer (10^{-4}–10^{-5} torr), and most preparative chemical vacuum systems are designed with this degree of vacuum in mind. Mercury exerts a pressure of approximately 10^{-3} torr at room temperature, so mercury vapor is present throughout a vacuum line which is equipped with mercury manometers and other mercury-containing apparatus. Since mercury vapor is condensable and relatively unreactive, its presence often can be ignored.

To start a vacuum system such as that illustrated in Fig. 5.1, the clean main trap is fitted in place with an even coat of stopcock grease on the joint. A Dewar partially filled with liquid nitrogen is raised around this trap, and the fore pump is immediately turned on. **CAUTION: Oxygen from the atmosphere will condense in a trap held at liquid nitrogen temperature; therefore, it is important never to leave a trap cooled to liquid nitrogen temperature exposed to the atmosphere for a significant length of time.** When the line has pumped down to less than 1 torr, the stopcocks are turned to route the gas through the diffusion pump, and the heater on the diffusion pump is then turned on. (If the diffusion pump is water cooled, the water should also be turned on.) The progress of the evacuation is followed by means of an electronic vacuum gauge or a McLeod gauge. If possible, the line should be evacuated for several hours before use (overnight is preferable for a large vacuum system) to permit the desorption of moisture and a thorough check on the performance of the line. If the line does not pump down rapidly, the most likely source of leaks is poorly greased stopcocks. Sometimes the offending stopcock or joint can be turned to work out the leakage path in the grease. Details on the proper greasing of stopcocks and joints are given in Chapter 8, and methods for hunting leaks are described in Chapter 7.

5.3 MANIPULATION OF VOLATILE LIQUIDS AND CONDENSABLE GASES

A. Freeze-Pump-Thaw Degassing. Liquids which have been exposed to the atmosphere or to inert gases contain appreciable amounts of dissolved gases that should be eliminated before use. If small quantities of liquid are involved, gases such as oxygen and nitrogen may be eliminated by means of the transfer process described in Section 5.3.C. In the more common situation of needing to degas a large volume of liquid, the freeze-pump-thaw technique is employed. This is accomplished by cooling the sample to liquid nitrogen temperature, pumping on the sample at this temperature, closing the stopcock or valve so the sample is isolated from the vacuum, allowing the sample to melt, then refreezing it and pumping away any noncondensable gases which boiled out of the material. This process is generally repeated several times. Note that ice and frozen methylene chloride, may break glass apparatus when they are thawed. The easiest course of prevention is to use a round-bottomed container which is one-fourth full or less. Another effective measure is to warm the tube rapidly so that a film of liquid forms between the container and the mass of the frozen solid. With experience, the latter technique can be effective in very difficult situations such as a water-filled thin-walled NMR tube.

B. The Transfer of Condensable Gases. Condensable gases are conveniently separated and transferred in a vacuum line by vaporization and condensation. Prior evacuation of the system and a suitably low temperature are necessary to quantitatively transfer the material. The vapor pressure corresponding to "quantitative" condensation of a compound will depend on the volume of the system and the quantity of material being transferred; as a general rule, a maximum vapor pressure of 10^{-3} torr is satisfactory. For simple transfer operations, liquid nitrogen (bp $-196°C$) is the most commonly used refrigerant.

C. Example: Transfer of a Condensable Gas. Suppose we wish to transfer the HCl from a 1-L glass bulb into a trap on the vacuum system (Fig.

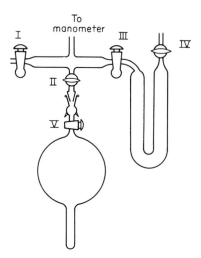

Fig. 5.5. The transfer of HCl from a bulb into a trap by condensation. See Section 5.3.C for a description of the transfer procedure.

5.5). The bulb is attached to the vacuum line by means of a carefully but lightly greased taper joint, and is held in place by springs, clamps, or rubber bands. The system, exclusive of the bulb, is evacuated to at least 10^{-3} torr pressure (stopcock V closed and II, III, and IV open). The U-trap is cooled with a Dewar of liquid nitrogen, and then HCl is admitted by slowly opening stopcock V. Stopcock IV is generally left open at this point because the HCl is quantitatively trapped at the boiling point of liquid nitrogen, $-196°C$, and the open stopcock allows any contaminating "noncondensable" gases such as oxygen and nitrogen to be removed. **CAUTION: Avoid pressurizing a vacuum line. Failure to keep the pressure below one atmosphere may result in stopcocks popping out with high velocity or explosion of the glass apparatus.** In the example just given, the bulb may have been of 1L capacity and filled with 1 atm of HCl. If this were condensed into a smaller U-trap, followed by closing stopcocks III and IV, the liquid nitrogen would have to be maintained around the trap to prevent the buildup of excessive pressure.

D. Trap-to-Trap Fractionation. Compounds of significantly different volatility can be separated by passing the vapors through a series of traps, each of which is cooled to a temperature which will condense one of the components. Judged by the standards of ordinary distillation, trap-to-trap distillation is very inefficient. Therefore the process is useful only when there is a large difference in the vapor pressures of the components being separated. A rough estimate of the separation efficiency may be gained from vapor pressure data (assuming the compounds do not strongly interact in the condensed state). If, at a particular temperature, the ratio of vapor pressures of two compounds is 10^4 or greater, an excellent separation should be possible; a ratio of 10^3 will usually lead to a good separation; and a ratio of 10^2 will permit fair separation. The separation is influenced by the rate of the trap-to-trap distillation: (1) the finite vapor pressure of the less volatile component will lead to its accumulation along with the more volatile component; and (2) if flow rates become too high and the more volatile component is in substantial excess, the more volatile gas will entrain the less volatile component and sweep some of it through the low-temperature trap. In order to achieve efficient separation, the first factor leads one to favor a quick distillation, while the second leads to the choice of a slow distillation.

Each separation problem presents its own peculiarities, so there is no single best procedure for trap-to-trap fractionation. For example, if the component of primary interest is a minor constituent mixed with a more volatile compound, the entrainment problem may be serious. To reduce entrainment, the mixture may be passed through several cold traps, and each trap should be deeply immersed in the cold bath to afford maximum contact with the cold surface. Sometimes entrainment is minimized by the inclusion of glass helices or indentations (Vigreaux) in the trap. However, a principal origin of entrainment appears to be the formation of microcrystallites in the gas phase and these are swept through the trap by the second component. Low flow rates and large-diameter tubing in the traps minimize this mechanical entrainment. To achieve a low flow rate, the volatilization of the mixture is often moderated by placing a chilled but empty Dewar around the sample tube as illustrated in the following example. The use of two or more traps cooled to the same temperature allows the crystallites to volatilize between traps and increases the chance that the component will condense on the walls of the next trap.

As already pointed out, it is possible to monitor the purity of the fractions by methods such as vapor pressure (see below) or gas phase infrared spectroscopy (Chapter 9). In planning an experiment, solvents and reactants are often chosen such that their vapor pressures facilitate the separation of reactants, products, and solvent. For compounds of similar volatility, more advanced separation techniques are required, such as fractional codistillation, low temperature column distillation, or gas chromatography, all of which are described in Chapter 9.

E. Example: Separation of BF₃ and CH₂Cl₂.

E. Example: Separation of BF_3 and CH_2Cl_2. Boron trifluoride (bp $-110.7°C$) is readily separated from methylene chloride (bp $40.7°C$) as illustrated in Fig. 5.6. Inspection of the vapor pressure data in Appendix V reveals that BF_3 exerts 75 torr at $-126°C$, whereas extrapolation of the vapor pressure data for CH_2Cl_2 to this temperature (log P vs. $1/T$ plot) indicates a vapor pressure of less than 10^{-3} torr for this component. Therefore, the reaction mixture is slowly passed through a trap cooled to $-126°C$ (methylcyclohexane slush bath, see below), which retains the methylene chloride, and into another trap at $-196°C$ (liquid nitrogen), which retains the boron trifluoride. The rate of trans-

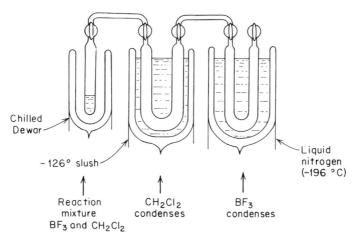

Fig. 5.6. Schematic representation of trap-to-trap fractionation for a BF_3-CH_2Cl_2 mixture. See Section 5.3.E for a description of the separation procedure.

fer of the gases through the traps is moderated by placing a chilled Dewar around the tube containing the reaction mixture. (This is accomplished by initially condensing the mixture in the left container at liquid nitrogen temperature, pouring the liquid nitrogen out of this Dewar, and replacing the chilled Dewar around the sample tube.) After the distillation is complete, the individual traps can be isolated to retain the components. The recovery of BF_3 should be quantitative. This may be verified by expanding the gas into a calibrated volume attached to a manometer, followed by pressure measurement and calculation of the moles present by the ideal gas law. Finally, purity checks may be performed by vapor pressure measurement, gas-phase infrared spectroscopy, and/or mass spectrometry. This sequence of separation, quantitative measurement of volatiles, and physical identification of components affords a powerful method of studying reactions involving at least one volatile substance.

F. PVT Measurement of Gases. One of the primary advantages of the vacuum line manipulation of gases is the ease with which quantitative measurements are made. If the problem simply requires dispensing measured amounts of gas, a procedure such as the following may be employed. The ideal gas law is sufficiently accurate for most chemical work if the compounds are well removed from their condensation temperatures and pressures. For example, the van der Waals equation of state for CO_2 indicates that 1 mmol of this gas in 25 mL will exert 749.7 torr pressure versus the ideal gas value of 748.4 torr. This disparity in pressures amounts to a 0.2% error, which is less that the other errors involved in routine PVT measurements, and is perfectly adequate for most chemical problems.

G. Example: Calibration of a Bulb and PVT Measurement of a Gas. In this example we are interested in obtaining a known quantity of gas, using as our gas source a storage bulb attached to the vacuum system. First it is necessary to have at hand a calibrated volume. If a bulb of known volume is not available, an evacuated gas bulb of the type shown in Fig. 5.5 is weighed, filled with water (if it is large) or mercury (if it is very small), and reweighed. The volume is then determined using the weight difference and the density of the liquid at the ambient temperature.

The clean, dry, calibrated bulb is attached to a vacuum system, such as site A on the line illustrated in Fig. 5.2, and evacuated to 10^{-3} torr or lower. Gas may be admitted to the calibrated bulb by isolating the working manifold from vacuum and opening the stopcocks and valves leading to one of the upper gas storage bulbs, C, while the pressure is monitored by manometer D. When the desired pressure is reached, the valve on the storage bulb is turned off, pressure and temperature measurements are made, and then the stopcock on the calibrated bulb is turned off. At this stage we know the pressure, temperature, and volume of the gas in the calibrated bulb, which permits the calculation of the number of moles via the ideal gas equation.

Before the gas in the calibrated bulb can be dispensed it is necessary to eliminate the gas which fills most of the working manifold and the manometer. This may be accomplished by condensing the excess gas back into the storage bulb and closing off that bulb. The gas which is contained in the calibrated bulb attached at site A can now be condensed into any of the evacuated traps on the line or into an evacuated reaction vessel, which might be attached at site B.

H. Example: Calibration of a Trap and Measurement of a Gas Sample. In contrast to the previous example, the objective here is to measure the amount of an entire gas sample, such as the BF_3 recovered in the trap-to-trap distillation discussed in Section 5.3.E. In this type of measurement a manometer is included in the calibrated volume, and it is necessary to account for the change of volume as the mercury level changes with changes in pressure. As described in Chapter 7 a constant-volume gas buret can be used, but this is somewhat cumbersome. So the simpler procedure described here is more frequently used.

Suppose that we wish to calibrate the volume of trap E connected to manometer D on the vacuum line in Fig. 5.2. A known quantity of a gas, such as CO_2, is condensed into trap E using the calibrated bulb and the techniques just outlined in Example 5.3.G. The stopcocks are then turned so that trap E communicates with the central manometer D but is isolated from the rest of the vacuum system. At this point, the cold trap is removed from E, the trap is allowed to come to room temperature, and the pressure and room temperature are measured. The volume of the manometer-trap combination is determined from the known moles of gas, the pressure, and the temperature, using the ideal gas law. The process is repeated with successively larger samples of CO_2, and a plot of volume versus pressure is constructed from the data. Since the bore of the manometer is of constant diameter, this plot should be a straight line. It also is possible to determine the variation of volume with pressure by previously measuring the inside diameter of the manometer tubing. The volume-versus-pressure plot is entered into the laboratory notebook for future use.

We can now determine the total moles of gas in a sample by condensing this gas into trap E, measuring the pressure on manometer D, measuring the room temperature, and using the volume-versus-pressure plot to determine the volume of the gas so that the ideal gas equation can be applied.

5.4 LOW-TEMPERATURE BATHS

A. Liquid Nitrogen and Liquid Air.   Liquid nitrogen is the most commonly used refrigerant for the manipulation of condensable materials on a vacuum system because it is easy to use, inexpensive, and will not support combustion. Furthermore, its boiling point is low enough, $-196°C$ ($77°K$), to lower the vapor pressure of most materials below 10^{-3} torr, which is necessary for the quantitative transfer of condensible materials on the vacuum system.

CAUTION: Skin contact with liquid nitrogen may lead to a frostbite burn. An occasional droplet of nitrogen, such as is encountered when filling a Dewar, often does not freeze the skin because of an insulating film of gaseous nitrogen which forms immediately. However, the skin is readily frozen if the liquid nitrogen is held on a spot by clothing which is saturated with the refrigerant, or by any other means which leads to extended contact.

CAUTION: The release of large quantities of liquid nitrogen in a confined space displaces air and can lead to asphyxiation. The areas where liquid nitrogen is dispensed or handled in significant quantities should be well ventilated.

In most large metropolitan areas liquid nitrogen is readily available commercially. It may be delivered as the liquid from the Dewar on a truck trailer to a large storage Dewar on site. More commonly, it is delivered in medium-size pressurized or atmospheric-pressure Dewars which can be moved into the laboratory (Fig. 5.7). As illustrated in that figure, liquid nitrogen is withdrawn from the pressurized Dewar by means of a valve and spigot, much as one draws water from a tap. The difference between filling a vessel with water or a small Dewar with liquid nitrogen is that the delivery tube and small Dewar must cool to $-196°C$ before liquid can be collected. Initially only cold gas will issue from the delivery tube. After the liquid starts to flow it will be rapidly volatilized in the small receiving Dewar until that Dewar is cooled to the boiling point of nitrogen. Similarly, liquid is generally drawn from medium-size atmospheric pressure Dewars by application of a small positive pressure of nitrogen gas (about 1–3 psig), which forces the liquid out of a dip-tube. Again there is an initial cool-down period as the dip-tube is initially inserted into the Dewar and the delivery tube and receiving Dewar are cooled.

The Dewars used on a vacuum system are generally of the wide-mouthed variety, with capacities ranging from 1/4 to 1 L, although larger Dewars are some-

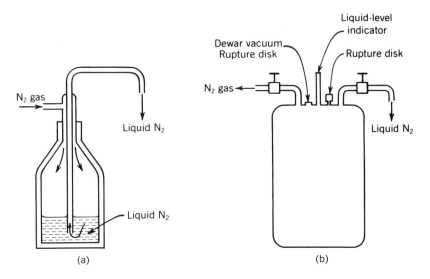

Fig. 5.7. Liquid nitrogen storage Dewars. (*a*) Atmospheric-pressure Dewar. Liquid nitrogen is drawn from the Dewar by applying a small positive pressure of nitrogen gas to the inlet on the sleeve of the metal dip-tube, forcing liquid nitrogen out the dip-tube. (*b*) Schematic drawing of a pressurized storage Dewar. Spigots are generally provided for both liquid and gas withdrawal.

times used on the main traps of a vacuum system. Because of the cooling time and the losses of nitrogen involved in drawing liquid nitrogen from a storage Dewar, it is common to employ a 3–5 L dispensing Dewar when the vacuum system is in operation (Fig. 5.8). This dispensing Dewar then requires only occasional filling, and it can be of a narrow-mouth design, which reduces evaporation losses. To further reduce losses, a loosely fitting cap or cork is generally placed over the mouth of a Dewar containing liquid nitrogen, and a Dewar which has a U-trap or storage tube protruding from its mouth is generally covered with glass wool. This loose cap also greatly reduces the condensation of oxygen (bp −183°C) in the Dewar. **CAUTION: It is very important never to seal a Dewar of liquid nitrogen, because the insulation on a Dewar is never perfect; so the liquid is constantly boiling and a sealed Dewar will eventually explode.**

Liquid air is a potential substitute for liquid nitrogen but is not commonly used because the vapors will support combustion and there is ordinarily little or no price advantage in its use.

B. Slush Baths. Liquid nitrogen and Dry Ice are convenient and inexpensive refrigerants. But, as shown in the examples in this chapter, a wider range of low-temperature baths is necessary for trap-to-trap fractionation of gases and for the characterization of a substance by vapor pressure measurements. A convenient constant-temperature slush bath consists of a mixture of a frozen compound in equilibrium with its liquid. The bath is made in a clean Dewar no more than

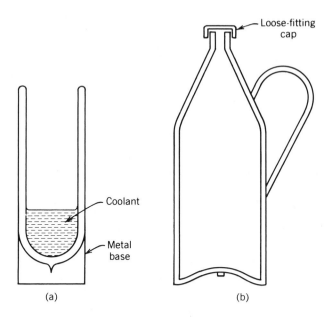

Fig. 5.8. Laboratory-size liquid nitrogen Dewars. (*a*) Cross section of a wide-mouthed glass Dewar. These range in capacity from ¼ to 1 L or larger, and are generally used to cool traps on vacuum systems. (*b*) Cross section of a metal Dewar. To minimize liquid nitrogen losses, a loose-fitting cap is usually placed over the mouth of the Dewar.

three-quarters full with the pure liquid and placed in a good hood. Freshly drawn liquid nitrogen is slowly added while the liquid is rapidly stirred with a stout stirring rod. There is a tendency for a hard frozen ring to form around the sides of the Dewar at the surface of the liquid. The addition of liquid nitrogen should be slow enough and the stirring adequate to minimize the size of this frozen ring, which in an extreme case will prevent the slush bath from fitting around a trap. Liquid air or liquid nitrogen which has condensed oxygen from the air should not be used to make slush baths of combustible compounds. The consistency of the final product should be similar to a thick malt. If the slush is too thick it cannot be penetrated by a trap, and if it is too thin it will not hold a constant temperature for long. A list of convenient materials for slush baths is given in Table 5.1.

Many slush baths have poor lasting properties, which makes it advantageous to precool the trap. This is accomplished by cooling the trap with liquid nitrogen, removing the Dewar of liquid nitrogen, and placing the slush bath *below* the trap. The stirring rod is then used to swab a spot on the trap with the slush. At first the bits of slush will freeze on the trap, but as the trap warms the frozen slush will melt. At this point the slush bath is raised around the trap. The liquids from a slush may be reused.

Table 5.1. Compounds for Cryostatic Baths

Temp, °C[a]	Compound	Temp, °C[a]	Compound
+ 6.55	Cyclohexane	−95.0	Toluene
+5.53	Benzene	−96.7	Methylene chloride
0.00	Water[b]	−111	Trichlorofluoro methane (Freon-11, bp 23.8°C)[c]
−8.6	Methyl salicylate	−111.95	Carbon disulfide
−15.2	Benzyl alcohol	−118.9	Ethyl bromide
−22.95	Carbon tetrachloride	−126.59	Methylcyclohexane
−30.82	Bromobenzene	−135	Dichlorofluoromethane (Genetron-21, bp 8.9°C)[c]
−37.4	Anisole	−138.3	Ethyl chloride
−45.2	Chlorobenzene		
−51.5	Ethyl malonate	−139	$CHCl_3$, 19.7% by weight C_2H_5Br, 44.9%
−57.5	Chloral		trans-$C_2H_2Cl_2$, 13.8% C_2HCl_3, 21.6
−63.5	Chloroform[d]		
−83.6	Ethyl acetate	−160.0	isoPentane

[a]Most of these values are taken from J. Timmermans, *Physico-chemical Constants of Pure Organic Compounds*, Vols. 1 and 2, Elsevier, Amsterdam, 1950 and 1965.
[b]Pure crushed ice in contact with distilled water.
[c]These fluorocarbons have low toxicity and are nonflammable. The liquid may be stored for reuse in a stoppered Dewar in a freezer.
[d]Reagent-grade chloroform contains a significant quantity of alcohol, which lowers the melting point.

C. Dry Ice Baths. In contrast to the fixed temperature of slush baths, a Dry Ice bath is somewhat less reliable because the equilibrium temperature is a strong function of the partial pressure of CO_2 above the bath. For example, one test in which Dry Ice was freshly powdered in air had a temperature 8°C below the normal sublimation temperature of −78.5°C. In another test, 10-h standing was required for a Dewar of Dry Ice to attain −78.5°C.[1] However, a small electric heater buried in the Dry Ice produced enough CO_2 to expel the air and attain the normal sublimation temperature in a matter of minutes. The temperature of an equilibrated Dry Ice bath at any barometric pressure may be found using the expression for sublimation pressure:

$$\log P = 9.81137 - \frac{1,349}{t + 273.16}$$

[1]R. B. Scott, in *Temperature, Its Measurement and Control*, Vol. 1, Reinhold, New York; 1941, p. 212.

where P is in torr and t in degrees Celsius. Generally Dry Ice is used in a bath containing acetone or another low-melting liquid to aid thermal conduction and promote equilibration.

5.5 VAPOR PRESSURE AS A CHARACTERISTIC PROPERTY

Vapor pressures, which are readily determined on a vacuum line, provide a convenient and sensitive means of identifying a volatile compound and/or checking its purity. The sample is condensed at liquid nitrogen temperature and all of the residual gases are pumped out of the system. If the compound has been exposed to the atmosphere immediately prior to the measurement, it should be degassed by several freeze-pump-thaw cycles.

The trap or tube attached to the vacuum system containing this sample is maintained at the desired low temperature by means of a slush bath, and the pressure is read on a mercury manometer. The sample must be large enough so that liquid is present while the measurement is being performed. The presence of an impurity of different volatility can be detected by the disparity between the known vapor pressure and the observed value. Another effective means of checking the purity is to remove by condensation a significant portion of the sample into a trap for temporary storage and then redetermine the vapor pressure. If the second vapor pressure is different from the first, impurities are present in the sample. Vapor pressure data for some common compounds at convenient slush bath temperatures are given in Appendix V.

5.6 MANIPULATION OF NONCONDENSABLE GASES

Liquid nitrogen is the coldest inexpensive refrigerant available. The two lower-temperature refrigerants, liquid hydrogen and liquid helium, are expensive, difficult to handle, and the former is dangerous because of the production of explosive vapor. Thus gases such as H_2, N_2, O_2, CO, and CH_4, which have high vapor pressures at liquid nitrogen temperature, $-196°C$, require special methods for their transfer. The two most commonly used techniques for these "permanent" gases are a reciprocating mercury pump (Toepler pump) and low-temperature adsorption on high-surface-area solids. The Toepler pump is very predictable and, when it is operating properly, quite trustworthy. However, it is expensive and complicated, and if mishandled it can fail spectacularly. Low-temperature adsorption is cheap, versatile, and easy to implement, but requires careful checking to avoid erroneous results.

A. Toepler Pump.   The principles of operation of a Toepler pump are outlined in Fig. 5.9, where it will be noted that a system of internal check valves permits gas from the vacuum system to enter the upper chamber, and subsequently to be compressed and expelled into a gas buret. The system may be man-

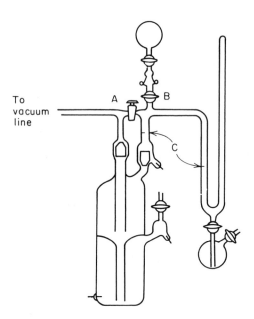

Fig. 5.9. Toepler pump and gas buret. In this installation a "constant-volume" manometer is used for measuring the pressure. Stopcock B allows the attachment of a sampling bulb or a large expansion bulb for use with large volumes of gas. Stopcock A is used for initial evacuation of the calibrated volume and sample bulbs. It is closed during the operation of the pump. When a gas is to be measured, the mercury levels in the right arm of the Toepler pump and in the left arm of the manometer are adjusted to levels (C) for which the gas volume has been calibrated. An excellent version of this pump is manufactured by the Rodder Instrument Co., Los Altos, Calif.

ual, but most often it is controlled by means of contacts which sense the mercury level. These are connected to a relay and solenoid valve which control the air pressure in the bottom chamber. On the inlet cycle, the gas being pumped is distributed between the vacuum system and the upper chamber of the pump. Thus the fraction of gas removed per cycle is equal to the volume of the upper Toepler pump chamber V_t divided by the volume of the vacuum system V_v plus the volume of the upper chamber. The fraction of gas remaining in the vacuum line after n cycles is

$$f = \left[1 - \frac{V_t}{V_v + V_t} \right]^n$$

from which it is easy to estimate the number of cycles necessary to collect the gas "quantitatively." The completeness of gas collection also can be judged by failure of gas to bubble through the outlet valve on the compression cycle, or by the lack of pressure change for the collected gas over several collection cycles.

The primary modes of failure of this type of pump are leakage past the internal check valves or some mishap which leads to a sudden rush of air into the upper or lower chamber. The first problem impairs the pumping action and is most commonly caused by poor check valve design or dirt on the check valves. The inrush of air from operator error or a malfunction results in a surge of mercury which may break the apparatus and create a shower of mercury.

B. Low-Temperature Adsorption. This method provides a convenient method of transferring and separating some permanent gases. The utility of low-temperature gas adsorption was recognized by Dewar in 1875, but the applications described here are in part based on the selectivity of certain solids which were not recognized until fairly recently.[2] Since the adsorption of one gas can affect that of others, materials which are condensable at $-196°C$ are first removed by the conventional condensation procedures discussed earlier in this chapter. The noncondensable gases are then adsorbed on a suitable solid surface at low temperature. Molecular sieve 4A at $-196°C$ will strongly adsorb H_2, O_2, N_2, CO, CH_4, and similar "permanent gases." Silica gel at this same temperature adsorbs hydrogen only lightly and the others strongly. Thus it is possible to separate hydrogen quantitatively from the other permanent gases by an adsorption trap of silica gel followed by another of molecular sieves; or, in place of the molecular sieves, one may use a Toepler pump to collect the hydrogen. It is important to cool the adsorbents before they are exposed to the gas mixture, because cooling silica gel in contact with hydrogen from room temperature to $-196°C$ leads to the retention of a small amount of the hydrogen in the silica gel. Heat flow is not efficient between the particles, so it is necessary to wait patiently for the trap to come to the desired temperature. In this connection, it is desirable to avoid traps with very large cross sections, which are especially slow to come to temperature.

This simple technique can be applied to a variety of transfer operations, and it often serves to replace the more cumbersome and expensive Toepler pump. The solid adsorbent should be sieved to eliminate fines or large particles. Particles of about 50 mesh (0.3-mm diameter) are ideal, and these should be contained in a suitable U-trap or other trap design with a small plug of glass wool on the inlet and outlet to prevent the adsorbent particles from being swept into the vacuum line (Fig. 5.10). Before use, the freshly packed trap is heated to about 300°C while being evacuated to a high vacuum. The initial evacuation and heating must be done slowly to avoid sweeping the adsorbent out of the trap by the copious quantities of adsorbed gases and water vapor which are evolved.

Low-temperature adsorption is also useful for the quantitative transfer of gases into special apparatus. For example, a small quantity of activated molecular sieve in a freeze-out tip on an infrared cell may be used for the identification of methane; or, when placed in a freeze-out tip on a manometer and calibrated volume, the molecular sieves permit the measurement of the quantity of noncondensable gas.

---

[2]The principles of selective adsorption by zeolites are discussed in D. W. Breck, *Zeolite Molecular Sieves*, Wiley, New York; 1974.

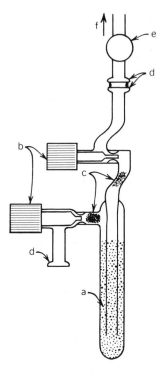

Fig. 5.10. Trap design for gas manipulation by low-temperature adsorption. (*a*) Solid adsorbent; (*b*) Teflon–glass valve (4 mm); (*c*) glass wool plugs; (*d*) O-ring connectors (9 mm); (*e*) ground-glass stopcock or Teflon–glass valve on high-vacuum line; (*f*) to high vacuum manifold. The trap may be moved from one attachment point on a high-vacuum line to another attachment point without exposing the adsorbent (and any gas sample collected in the trap) to atmospheric gases.

5.7 SAFETY

As with most scientific apparatus, improper handling of the vacuum line can be hazardous. Appendix I presents a detailed listing of laboratory hazards, and Section 7.1.E describes the correct handling of mercury. This brief section will serve as a reminder of potential safety problems which are encountered in the use of a vacuum system. Since there always is the possibility of explosion or implosion of pressurized or evacuated glassware, eye protection should be worn at all times around a vacuum line. Other hazards include: explosion of glassware which is inadvertently pressurized by the uncontrolled warming of condensed gases in a small volume; implosion of evacuated storage bulbs and Dewars (these items should be surrounded with tape or a special plastic coating to contain flying glass); exposure to mercury, which is a cumulative poison (spills should be promptly and thoroughly cleaned up); asphyxiation by nitrogen gas (liquid nitrogen should not be handled in a small, poorly ventilated room); and frostbite from the contact of skin with low-temperature refrigerants.

CAUTION: Since electricity is used to run equipment such as vacuum pumps, gauges, and Tesla coils, proper care must be exercised when using and maintaining this equipment. Individual items should be grounded through a three-prong plug, and vacuum racks must be properly grounded, for example, by means of a copper wire between the vacuum rack and a water pipe.

GENERAL REFERENCES

Angelici, R. J., 1977, *Synthesis and Technique in Inorganic Chemistry*, 2nd ed., Saunders, Philadelphia. This book provides a useful exercise in vacuum line practice for the newcomer to the field.

Brauer, G., Ed., 1981, *Handbuch der Preparativen Anorganischen Chemie*, 3rd ed., Enke Verlag, Stuttgart. The first chapter includes some information on vacuum line technique, and various syntheses throughout this three-volume set illustrate the application of vacuum systems to inorganic preparations. The English edition of the second edition of this book is of some use, but it is not up-to-date.

Jolly, W. L., 1970, *The Synthesis and Characterization of Inorganic Compounds*, Prentice-Hall, Englewood Cliffs, N. J. An excellent, concise introduction to chemical vacuum line technique, with some experiments which will provide good exercises to the newcomer to chemical vacuum lines.

Melville, H., and B. G. Gowenlock, 1964, *Experimental Methods in Gas Reactions*, Macmillan, London. Primarily devoted to experimental methods for gas-phase kinetic studies, but the principles and apparatus are of interest in general chemical vacuum line technology.

Moore, J. H., C. C. Davis, and M. A. Coplan, 1983, *Building Scientific Apparatus*, Addison-Wesley, Reading, Mass., A good, physically oriented reference on laboratory technique. Chapters 3 and 4 cover glassblowing and vacuum apparatus, respectively.

O'Hanlon, J. A., 1980, *A User's Guide to Vacuum Technology*, Wiley, New York. A practical guide to the general area of vacuum technology and related primarily to semiconductor and optics technology.

Sanderson, R. T., 1948, *Vacuum Manipulation of Volatile Compounds*, Wiley, New York. A good but somewhat out-of-date discussion of vacuum line techniques.

Stock, A., 1964, *Hydrides of Boron and Silicon*, Cornell University Press, Ithaca, N.Y. In connection with this presentation of his pioneering research on boron hydrides, Stock describes the chemical vacuum line techniques which he invented.

6

Pumps for Rough and High Vacuum

This chapter is primarily devoted to pumps for high vacuum-operation (10^{-3}–10^{-5} torr), which is the vacuum range of greatest interest in chemical vacuum lines. In addition, rough-vacuum systems (760–0.1 torr) are discussed in connection with their use in manipulating mercury-filled apparatus, such as Toepler pumps and McLeod gauges.

6.1 ROUGH-VACUUM SYSTEMS

A. Applications for Rough Vacuum. In the description of the operation of a Toepler pump given in Chapter 5, mercury was transferred from a lower reservoir to an upper pumping chamber by the use of pressure differentials. The upper chamber communicates with the high-vacuum line and the lower reservoir with a rough-vacuum system or the atmosphere. Mercury is pulled from the upper chamber into the lower reservoir by means of a rough vacuum applied to the lower reservoir, and on the other stroke, mercury is forced into the upper chamber by admitting air to the lower reservoir. There are other types of equipment where mercury is added to or withdrawn from the vacuum system by similar means, including the constant-volume gas buret (Chapter 7) and various tensimeters (Chapter 9).

B. Design of Equipment. Since the rough-vacuum system may have a practical vacuum limit of 1 torr to several tens of torr, equipment must be designed and filled with mercury to take into account the disparity in pressures between the high-vacuum side and the mercury reservoir. The reservoir must be placed low enough so that the apparatus can function properly with a differential in the mercury heights on the high-vacuum and reservoir sides. It is also important to bear this differential in mind when filling apparatus with mercury.

C. Rough-Vacuum System. A relatively inexpensive single-stage pump (see Section 6.2) will serve well for most rough-vacuum applications. It is important to protect this pump, by means of an uncooled trap of appropriate size, from the possible inrush of liquid mercury which may occur during an accident. To avoid the buildup of toxic mercury fumes in the work area, the exhaust from the pump is conducted to a hood. The rough-vacuum manifold may be constructed from glass, copper, or polyethylene tubing. The latter, more robust materials are often preferred because the rough-vacuum manifold often is placed below the main vacuum system and is thus vulnerable to items that are dropped. A typical rough-vacuum system is illustrated in Fig. 6.1.

6.2 HIGH-VACUUM PUMPING SYSTEMS

A. Rotary Oil-Sealed Pump Designs. All of the commercial oil-sealed mechanical pumps are based on a rotor inside a cylindrical stator. These individ-

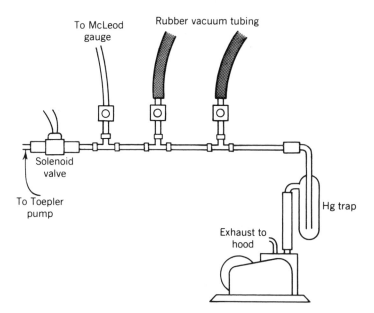

Fig. 6.1. Rough-vacuum system. Frequently used items such as the Toepler pump and constant-volume manometer are often connected permanently into the rough-vacuum system. Also, one or two outlets, with vacuum tubing attached, are included for general use. The manifold generally is constructed from rigid plastic or metal tubing.

ual stages will be discussed below. Of greatest importance is not the design of these stages, but rather whether the pump has a single stage or two stages connected in series. A well-constructed single-stage pump is capable of attaining about 0.01 torr, but a more practical limit is on the order of 1 torr. This type of pump is useful in rough-vacuum systems. Two-stage pumps generally are quoted to attain between 10^{-3} and 10^{-4} torr, although pumping speeds are slow near these pressure limits. Two-stage pumps are generally preferred for backing diffusion pumps in high-vacuum systems.

There are three common designs for the internal workings of rotary oil pumps and these are illustrated in Fig. 6.2. All employ a rotor which revolves inside a cylindrical stator; a seal between the fixed and moving parts is maintained by a thin film of oil. Of the three designs, the internal-vane pump is by far the most common in the laboratory; it is generally used to back a diffusion pump in high-vacuum systems. Internal-vane pumps generally make relatively little noise when operating at high vacuum. The rugged and easily repaired external-vane pump is mainly available in small pump sizes and finds its greatest use in rough pumps. These pumps appear to stand up to a hostile environment better than the internal-vane pumps, but they have the disadvantage of producing a loud clanking sound when operating at their vacuum limit. The rotary-plunger pumps are used in high-vacuum applications and are fairly common in large vacuum systems. Very good pumps are available in all three designs.

For many years the connection between the pump and motor was made with a V-belt, but "direct-drive" pumps are becoming common. The latter are more compact, smoother running, and perhaps a bit quieter and more reliable. On the other hand, the semiflexible coupler between the motor and pump shaft can

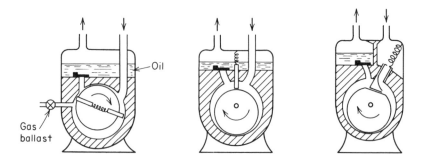

Fig. 6.2. Internal-vane, external-vane, and plunger-type rotary vacuum pumps. It will be noted that the internal-vane pump involves a rotor concentric with the drive shaft, but which is off-center with respect to the stator. By contrast, the external-vane and rotary-plunger pumps have a rotor which is asymmetric with respect to the shaft; however, the shaft is centered in the stator. All three involve close tolerances, so the high-vacuum performance is impaired by particles of dirt or corrosive gases. Some pumps are partially constructed from soft die-cast metal, which is eroded by mercury. Special corrosion-resistant pumps and inert flourinated pumps are used in the semiconductor industry.

wear out, and when this does happen, the coupler is more difficult to replace than a belt.

Many vacuum pumps are supplied with a gas ballast valve which is designed to prevent the condensation of moisture and similar condensable materials. This valve bleeds a small amount of air into the pump and as a result the ultimate vacuum and pumping speed of the pump suffer. There is no occasion to use this feature on a pump on a typical chemical high-vacuum line, and it rarely, if ever, is used in typical laboratory rough-pumping systems.

B. Troubleshooting and Mounting a Mechanical Pump.   It will be noted in Fig. 6.2 that gas from a rotary pump is exhausted through an immersed flapper valve. For this valve to make a good seal it must be covered with oil. One common cause of the loss of ultimate vacuum performance in these types of pumps is a low oil level, and this should be the first item to be checked when a pump is not performing well. A telltale sign of low oil is a change in the sound of the pump.

With a little experience the experimentalist becomes used to the sound of a properly operating pump, and any deviation in this sound is a sign of trouble. The distinctive sound when there is a small leak in the system, or a low oil level (the sounds are often similar), or the slapping sound characteristic of a worn belt, is a tip-off that remedial action is necessary before the problem gets worse.

A vacuum pump should be scrupulously protected from corrosive vapors and materials which will be absorbed in the pump oil or condense in the pump. For most laboratory operations a low-temperature trap is employed for this purpose, and in the case of fluorine handling systems a soda-lime trap is used to neutralize the corrosive gases. Despite these precautions, the pump oil does eventually break down and become contaminated. Regular oil changes should be scheduled for a pump at about yearly intervals for a well-protected pump and more often for pumps which are not well protected.

Special corrosion-resistant pumps are now widely used in the semiconductor industry. Often these are equipped with pump oil purifiers and are never turned off when contaminants are present. These pumps are expensive and under harsh conditions they break down much more often than a well-protected conventional pump.

As mentioned in Section 5.1.B, the transmission of vibration from the mechanical pump to a vacuum rack decreases the accuracy of manometer readings. To minimize this problem, it is generally best to mount the pump on the floor below a vacuum rack and to connect the pump to the system with a short length of heavy-walled rubber tubing, Tygon tubing, or flexible metal tubing.

C. Diffusion Pump Design and Operation.   The pumping speed of a two-stage mechanical pump generally drops very rapidly below 10^{-3} torr. By contrast, Fig. 6.3 shows that a typical diffusion pump has good pumping speed below this pressure. Therefore, a tandem arrangement of a diffusion pump with a

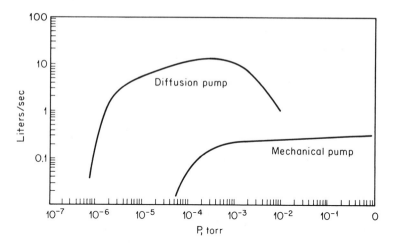

Fig. 6.3. Comparison of the pumping speed of a typical two-stage mechanical pump with a single-stage diffusion pump.

rotary fore pump is commonly employed for the evacuation of high-vacuum systems.

The design of a diffusion pump is illustrated in Fig. 6.4. These pumps operate by boiling mercury or a low-vapor-pressure oil and conducting the vapors out of a nozzle. This nearly unidirectional stream of vapor molecules collides with gas molecules which have diffused into the pump from the system being evacuated. As a result of these collisions, momentum is imparted to the gas molecules in the direction of pumping. The vapor is then condensed and returned to the boiler, and unwanted effluent gas is removed by the mechanical fore pump. Frequently the diffusion pump is designed with two or more nozzles in series to allow operation of the pump at a high fore pressure without sacrificing ultimate vacuum. An oil pump is faster than a mercury pump of comparable size because of the higher collisional cross section of the large oil molecules and the higher velocities of these higher boiling materials. An oil diffusion pump requires periodic replacement of the fluid, whereas a mercury pump can operate many years without attention. Among the commonly used fluids for diffusion pumps presented in Table 6.1, the silicone oils are the most popular in chemical systems. Because of increased awareness of the health hazards of mercury, oil diffusion pumps have become the standard in most laboratories. When a mercury diffusion pump is employed, it is highly advisable to include a cold trap ($-78°C$ is satisfactory) between the diffusion pump and the fore pump to reduce the amount of mercury vapor reaching the atmosphere.

The rate of boiling of the pumping fluid is determined by trial and error. The use of too low a rate of boiling will reduce both the pumping speed and the limiting fore pressure. In general, the heater input should be high enough so that the

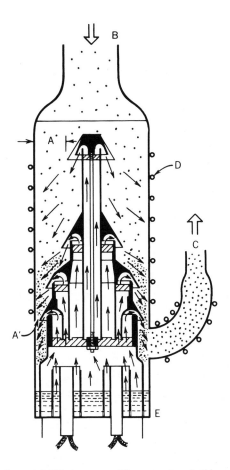

Fig. 6.4. Cross-section of a metal diffusion pump. The upper stage in this pump has a wide annular opening (A) which provides a good ultimate vacuum. The lower stage has a small annular opening (A') so the pump will operate against a high fore pressure. (B) High-vacuum connection to the low-temperature trap and vacuum line. (C) Connection to rotary oil-sealed pump. This pump is cooled by means of water tubes (D). Air-cooled versions have fins in place of these tubes and a fan is installed to blow air over these fins. (E) Electrically heated oil reservoir.

vapor is condensing below the nozzle at a brisk rate. If the boiling rate is too fast, the pumping efficiency goes down because of backstreaming of the pumping vapors. Bumping, or eruptive boiling of the fluid, has an adverse effect on the pumping action and is to be avoided. The source of bumping is an inversion in the vapor pressure of the fluid, owing to cooling of the surface of the fluid by evaporation while heat is applied from below. Because of its low thermal conductivity and high heat of vaporization, oil is much more susceptible to bumping than is mercury. Heaters which protrude into the pumping fluid are generally most effective in reducing bumping.

Table 6.1. Comparison of Fluids for Diffusion Pumps

Fluid	Type	Vapor pressure at 25°C (mm)	Resistance to Air Oxidation
Mercury	Metal	2×10^{-3}	Best
Apiezon A	Hydrocarbon	2×10^{-5}	Poor
Apiezon B	Hydrocarbon	4×10^{-7}	Poor
Apiezon C	Hydrocarbon	10^{-8}	Poor
Silicone 702	Silicone	—	Better
Silicone 704	Silicone	$10^{-6} - 10^{-8}$	Better
Silicone 705	Silicone	$10^{-9} - 10^{-10}$	Better
Octoil	Ester	2×10^{-4}	Fair
Octoil S	Ester	2×10^{-5}	Fair

D. Matching the Fore Pump and Diffusion Pump. A diffusion pump has a critical pressure above which it does not operate. Therefore, the fore pump must be capable of achieving this limiting pressure in a reasonable time period. Furthermore, the fore pump should be capable of removing at least as many moles of gas per unit time as the diffusion pump delivers. The throughput (speed in volume per unit time $\times$ pressure) is proportional to the speed on a molar basis; so a match of the pumps is obtained when $S_1 P_1 = S_2 P_2$. In this equation, S_2 is the speed of the diffusion pump at the appropriate high vacuum P_2, P_1 is the limiting fore pressure for the diffusion pump, and S_1 is the pumping speed of the mechanical pump at this pressure. Generally, the pump manufacturers provide the information necessary for approximate throughput calculations.

E. Cold Traps. In addition to the fore pump and the diffusion pump, a third active pumping element in most chemical vacuum systems is a liquid nitrogen-cooled trap situated between the main high-vacuum manifold and the diffusion pump. This trap pumps the system by removing condensables, and it serves to protect the diffusion and mechanical pumps from corrosive vapors. It is undesirable to accumulate large quantities of condensables in this trap and it is poor practice to use the trap to discard unwanted condensables in the vacuum system. (These unwanted materials should be condensed into a tube attached to the working manifold and removed to a hood.) Even a slight buildup of condensables eventually causes some loss of vacuum. As the liquid nitrogen level falls, the material condensed near the top of the trap will volatilize; while most of it is immediately recondensed in the lower portion of the trap, a small amount diffuses back into the vacuum line. Thus, when the vacuum is critical, the Dewar should be kept full of liquid nitrogen. Special trap designs prevent this problem by condensing the vapors at the bottom of the liquid nitrogen reservoir rather than the top.[1] However, these traps are cumbersome and difficult to clean, so they are not commonly used on chemical vacuum lines.

A trap design which allows periodic cleaning is highly desirable for chemical vacuum systems. Two such designs are illustrated in Fig. 6.5, where some of the desirable features are described.

F. Arrangement of the Pumping Components. To avoid oxidation of the pumping fluid, the hot diffusion pump should not be exposed to large quantities of air. Therefore, an arrangement of stopcocks is used to bypass the diffusion pump when large amounts of air are being pumped. This bypass feature also makes it unnecessary to wait for the diffusion pump to cool down when the main trap must be removed for cleaning or when the vacuum system is turned off

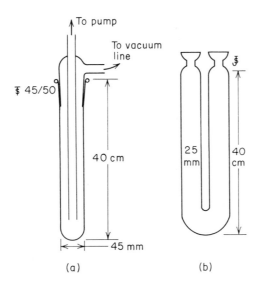

(a)   (b)

Fig. 6.5. Main traps. (a) This trap is long enough to extend to the bottom of a 1-L Dewar and sufficiently small in diameter to avoid excessive displacement of liquid nitrogen. These features allow the trap to remain partially immersed in liquid nitrogen after standing overnight. The arrangement of inlet and outlet shown here minimizes clogging of the trap because most of the condensate is deposited on the large outer tube rather than the smaller inner tube. (b) A U-trap is advantageous if there is a likelihood of condensing significant quantities of pyrophoric or highly toxic compounds, because a finite amount of material is always condensed on the center tube of trap (a). To remove a trap such as (b) from the line, one should fill it with nitrogen and quickly take it to a hood, where a slow stream of nitrogen may be passed through it to moderate the reaction which will occur when the contents of the trap warm up and volatilize. Ball joints facilitate the removal of a U-trap.

[1]A. Jordan, *J. Sci. Instr.*, 39, 447 (1962); H. W. Jones, *Rev. Sci. Instr.*, 35, 1240 (1964). Traps of this type, having internal liquid nitrogen reservoirs, are available from Eck & Krebs Scientific Laboratory Glass Apparatus Inc., 27-09 40th Ave., Long Island City, NY 11101.

and the fore pump is vented to the atmosphere. This bypass may be accomplished with one two-way stopcock in conjunction with a conventional off-on stopcock, as illustrated in Fig. 6.6. Also, a special three-way stopcock is available which combines these two functions.[2] To maintain good pumping speed in the system, large-diameter tubing and large stopcocks should be used on items leading into the diffusion pump. Items connecting the diffusion pump outlet and the mechanical fore pump can be of smaller diameter because the pressure is higher in this section.

G. Example: Initial Pump-Down and Venting the Pumping System.
Initial evacuation of the system is accomplished by attaching all

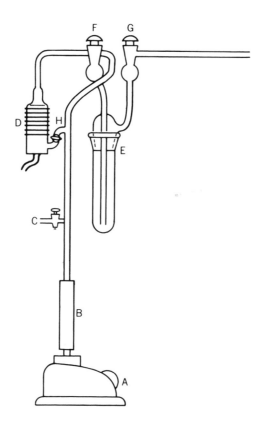

Fig. 6.6. Pumps, traps, and bypass. (A) Mechanical fore pump; (B) short length of vacuum tubing; (C) stopcock for venting system when fore pump is turned off; (D) diffusion pump; (E) cold trap at $-196°C$; (F), (H) stopcock which allow the diffusion pump to be isolated and bypassed; (G) stopcock which allows isolation of the main manifold from the trap and pumps.

[2]Item 8555 from Eck & Krebs (see footnote 1).

traps, starting the fore pump, and evacuating the vacuum line. At this point a Dewar containing liquid nitrogen is slowly raised around the main trap (E in Fig. 6.6), the stopcocks are turned to route the gases from the main trap (E) through the diffusion pump (D), and on to the fore pump. The diffusion pump is turned on and the system allowed to pump down.

To shut the system off, stopcocks (F) and (H) are turned to isolate the diffusion pump and to establish straight communication between the main trap (E), and the fore pump; also, the main valve to the vacuum system (G) is closed. The following set of operations is then carried out in sequence and without delays between each step: the fore pump is switched off, stopcock (C) is opened to bring the system to atmospheric pressure, the liquid nitrogen-filled Dewar on trap (E) is lowered, and trap (E) is removed and placed in a hood. The diffusion pump heater can then be switched off. The reason for carrying these steps out in fairly rapid sequence is that once the fore pump is turned off, oil may back up into the evacuated system if the flapper valve in the pump fails. In addition, if trap (E) is left at liquid nitrogen temperature for very long, it will condense oxygen from the atmosphere; so it is important to remove the liquid nitrogen and discard the contents of the trap soon after the system is vented. **CAUTION: It is important never to leave a trap which is cooled by liquid nitrogen open to the atmosphere. The condensed liquid oxygen presents an explosion hazard in contact with oxidizable materials and sudden warming or bumping of an oxygen-filled trap will pressurize the system, with potential breakage or explosion.**

If pyrophoric materials are present in trap (E), it may be safest to isolate this trap from either pump by stopcock (F), and carry out a trap-to-trap distillation of the material into a tube attached to the vacuum system. If this tube is equipped with a stopcock, it can later be vented into a nitrogen stream in a hood. Alternatively, a U-trap of the type shown in Fig. 6.5 may be used and the procedures described in that figure caption may be followed.

H. Limitation of Pumping Speed by the Apparatus.   The time required to reduce the vacuum in a line below 10^{-3} torr is often limited by the stopcocks and tubing in the line and not by the pumps. In this situation, little advantage is gained by the use of a large fore pump-diffusion pump combination, and, as described previously, the increased vibration of a larger mechanical pump may lead to inaccurate manometer readings. The major limitations on pumping speed occur at pressures where the mean free path of a molecule exceeds the small dimensions of the system. Under these conditions collisions with the walls become more frequent than intermolecular collisions. For example, the mean free path of oxygen at 10^{-3} torr and 25°C is about 5 cm, which is much larger than the usual tubing diameter.

It is instructive to consider the formula for the number of molecules per second, q, streaming from one end of a cylindrical tube of diameter d and length L to the other, Eq. (1).

$$q = \frac{Nd^3\pi}{3L\sqrt{2\pi MR}} \left(\frac{P_2}{\sqrt{T_2}} - \frac{P_1}{\sqrt{T_1}} \right) \tag{1}$$

P_1 and T_1 are the pressure and temperature at one end and P_2 and T_2 refer to the other end, M is the molecular weight of the gas, R is the gas constant, and N is Avogardo's number. This equation applies in the above-mentioned (molecular flow) region, which commences at about 10^{-2} torr. Of particular importance is the direct proportionality of the flow to the cube of the tube diameter. Thus, large-diameter tubing and large-bore stopcocks improve the pumping speed at low pressures. As with a series electrical circuit, the total impedance (proportional to $1/q$) is equal to the sum of the individual impedances, Eq. (2).

$$\frac{1}{q_{total}} = \frac{1}{q_1} + \frac{1}{q_2} \tag{2}$$

Therefore, one small-bore stopcock or a length of small-bore tubing will predominate in the determination of the impedance of the system. The analogy with electrical circuitry also holds for the parallel arrangement of impedances. Thus, if there are two or more connections between a manifold and the source of vacuum, the impedance q is given by Eq. (3).

$$q = q_1 + q_2 \tag{3}$$

To take advantage of this effect, and also to be able to pump several sections of the line at once, it is advantageous to design the main high-vacuum manifold (e.g., Fig. 5.2) and the main trap with larger-diameter tubing and larger stopcocks than are found in the working manifolds.

GENERAL REFERENCES

There are many good books on the scientific and technical aspects of vacuum and only a few are cited here.

Duschman, S., 1962, *Scientific Foundations of Vacuum Technique*, 2nd ed., rev. by J. M. Lafferty et al., Wiley, New York. Pumps for high vacuum and ultrahigh vacuum are discussed in some detail, and equations for the kinetic behavior of gases are presented.

Holland, L., W. Steckelmacher, and J. Yarwood, 1974, *Vacuum Manual*, E. and F. N. Spon, London (in the U.S.: Halsted Press div. of John Wiley & Sons Inc.). A very useful general reference, covering principles of gas flow, materials of construction, sealants, and commercial equipment.

Roth, A., 1982, *Vacuum Technology*, 2nd ed., North-Holland, Amsterdam. This excellent, comprehensive book covers similar material to the above volumes and it does not neglect rough-vacuum and high-vacuum regions.

Wissler, G. L. and R. W. Carlson, Eds., 1979, *Vacuum Science and Technology*, Vol. 14 of *Methods of Experimental Physics*, L. Marton and C. Marton, Eds. Academic, New York. An excellent multiauthor volume covering all major aspects of vacuum technology. The emphasis is on physical application and ultrahigh vacuum, but there is plenty of useful information for chemical applications.

O'Hanlon, J. F., 1980, *A Users Guide to Vacuum Technology*, Wiley, New York. This book is oriented somewhat to semiconductor and optics applications. It has very useful sections on residual gas analysis.

7

PRESSURE AND FLOW
MEASUREMENT AND
LEAK DETECTION

The measurement of pressures is quite important in vacuum line work, where determinations are performed routinely in the medium- or high-vacuum range to estimate the initial vacuum in a system, and in the 1–760-torr range to measure gases which are being manipulated. Flow measurement and control is commonly employed in catalytic flow reactors, in reactors for the growth of electrical and optical materials from gas-phase reagents, and in gas chromatographic analytical systems. Leak detection is related to these topics and is of major importance in troubleshooting vacuum and gas-handling systems. The implementation of these pressure and flow measurements and techniques for leak testing will be covered in this chapter.

7.1 MANOMETRY (1–760 TORR)

A. Mercury Manometers for Routine Work. The pressure-volume-temperature measurement of gases is the backbone of quantitative chemical vacuum line work. For these measurements an error of a few percent is frequently sufficient and may be attained by a simple U-manometer attached to a calibrated volume and read with an "inexpensive" cathetometer.

The manometers illustrated in Figs. 7.1 and 7.2 are convenient for use in conjunction with a chemical vacuum system. The simple U-manometer in Fig. 7.1 minimizes the amount of mercury used and therefore minimizes the mass of the manometer. The bubbler manometer illustrated in Fig. 7.2 uses more mer-

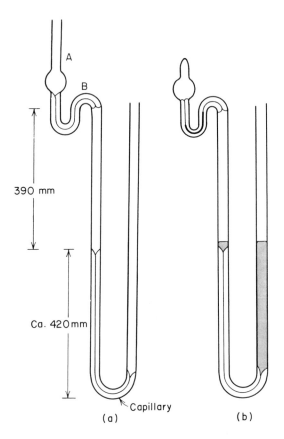

Fig. 7.1. The dimensions for a U-manometer. (*a*) U-manometer before it is filled with mercury. The dimensions shown give a range of pressures up to 1 atm. Using capillary tubing in the lower part of the manometer reduces the volume of mercury required. (*b*) Mercury-filled U-manometer. The procedure for filling the manometer is as follows. First, triple-distilled mercury is filtered directly into the manometer until the mercury column is about 425 mm high in both legs (i.e., about 5–10 mm above the union with the capillary tubing in the reference leg). Both sides of the manometer are then evacuated and gently flame-dried. The evacuated reference side is sealed off at point A, and mercury is forced into the small upper U by pressurizing the measuring arm. This traps any residual gas in the small upper chamber. To retain this residual gas permanently, a vacuum is established in the reference arm by gentle tapping or heating at point B while evacuating the measuring arm. This final process may be repeated on occasion to ensure a good reference vacuum.

cury but provides pressure release, and is therefore desirable for applications where excess pressure might build up inadvertently, as, for example, in the introduction of gases from a compressed-gas cylinder or in chemical operations with low-boiling solvents, such as liquid ammonia. Filling instructions for both manometers are given in their respective figure captions. Accurate manometry requires a clean mercury surface, so either manometer should be filled with freshly filtered triple-distilled mercury, and the interior of the manometer

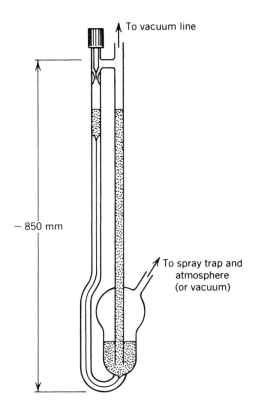

To vacuum line

~ 850 mm

To spray trap and
atmosphere
(or vacuum)

Fig. 7.2. A versatile bubbler manometer. The bubbler manometer is securely mounted by the reservoir and attached to the vacuum system. It is then easily filled by the following process. The level of the bottom end of the vertical tube dipping into the reservoir is marked on the outside of the reservoir. Next, a calculated amount of mercury is filtered into the reservoir. With the valve between the two arms open, a vacuum is *slowly* drawn on the manometer. The mercury level must not drop below the mark on the reservoir, or else bubbles will enter the vertical tube and shoot mercury through the vacuum system. If the mercury level in the reservoir comes close to the mark, the manometer is brought up to atmospheric pressure and more mercury is added. When the proper amount of mercury is present in the fully evacuated manometer, the mercury level should be about 10 mm above the mark on the reservoir, and the upper meniscus should be in a region of the manometer suitable for measurement, as illustrated. Once the manometer is properly filled and evacuated, the valve is closed to isolate the reference arm at high vacuum.

should be clean and dust-free. The proportions of these manometers and the amount of mercury should be arranged such that pressures from zero to somewhat over 760 torr can be spanned by the manometer without gas being forced around the bottom of the U-manometer or out of the bubbler on the bubbler manometer. To minimize errors arising from capillary depression, the two arms of the manometer should be of the same diameter in the region of the menisci (a minimum internal diameter of 8 mm is preferable).

B. Manometric Measurements. The ideal gas law is generally sufficiently accurate to measure the amount of gas at the subatmospheric pressures encountered in vacuum line work. When the gas is being measured in a U-trap with an attached manometer, the volume of the system will change with pressure because of the varying heights of the mercury level. One way of avoiding this complication is to employ a manometer in which the mercury level can be adjusted to a fixed height in the measuring arm by introduction of mercury from a reservoir. However, this is cumbersome, so the usual practice is to calibrate one leg of a simple manometer (Fig. 7.1 or 7.2). This process has been described in Section 5.3.H.

To achieve acceptable errors arising from capillary depression of the mercury, the manometer arms should be constructed from glass tubing having the same internal diameter, preferably not less than 8 mm. When the interior surface of the manometer adsorbs vapors or becomes dirty, the meniscus on the exposed leg of the manometer often will not have the same height as that on the reference side. If this condition arises and if small pressures are being measured, it may be necessary to make a capillary depression correction as indicated in the next section. Such corrections are unnecessary in routine work (i.e., in the 100–700 torr pressure region) using a clean manometer.

Ordinarily, the largest error in manometery comes in measuring the differential mercury column height. A cathetometer (Fig. 7.3), is routinely used for this process, but a meter stick attached to the manometer provides sufficient accuracy for rough measurements. The cathetometer consists of a telescope with a crosshair which is aligned with the top of the mercury meniscus. This telescope is mounted on a calibrated vertical scale equipped with a supplementary vernier scale for accurate interpolation of the readings. Before use, the cathetometer must be leveled so the travel of the telescope is indeed vertical. Similarly, if a manometer is read by means of an attached meter stick, the manometer must be strictly vertical. This can be accomplished at the time of construction by using a bubble level or a plumb line. It is good practice to mount all manometers on a true vertical.

Whether the meniscus height is measured with a cathetometer or a meter stick, it is important to have good illumination of the meniscus and to do the illumination in a consistent manner throughout a set of measurements. A sheet of white paper behind the manometer often provides good definition of the meniscus. Under many lighting conditions, illumination of this paper from behind may further improve the visibility of the meniscus.

C. Manometers for High Accuracy Work. For a mercury manometer with one arm evacuated, the pressure is given by $P = hdg$, where h is the difference in the height of the mercury columns, d is the density of mercury, and g is the acceleration due to gravity. In the old set of pressure units the pressure was reported in mm of Hg. This pressure unit was defined for mercury at $0°C$, where its density is 13.5951 g/cm^3, and the acceleration due to gravity is $g_0 = 980.665$

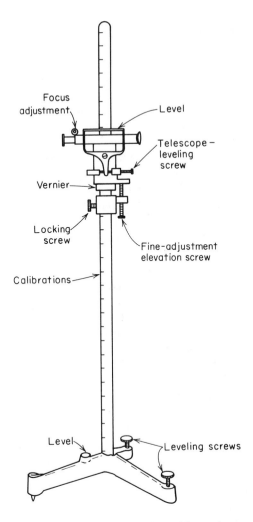

Fig. 7.3. Cathetometer. This type of cathetometer is used in routine manometry. It is light in weight and may be moved easily from one site to the next. A sturdy three-legged platform which is the same height as the vacuum line bench may be used to bring the cathetometer to the proper height for the observation of manometers on the vacuum line.

cm/sec^2. More recently, the torr pressure unit has come into wide use, and to within 7×10^{-4} percent the torr is equal to the mm of Hg. (Further definitions of the various pressure units may be found in Appendix VI.) Precise work requires the correction of the mercury height by a factor of g/g_0 (where g is the local value for the acceleration due to gravity), and a second factor of d_t/d_0, which is the density of mercury at the measuring temperature divided by the density at 0°C. Values for the ratio d_t/d_0 are given in Appendix VI, and the local value of g/g_0 may be determined from the data given in the *CRC Handbook of Chemistry and Physics*.

If the manometer arms are not of equal diameter or if their interior surfaces are covered by different adsorbed molecules, the menisci in the two arms will be different and a capillary depression error is introduced. The accuracy of this correction is low, so it is best minimized by employing large-bore tubing and the purest available mercury. Despite these precautions the menisci are rarely of identical height. A table of capillary depression corrections is given in Appendix VI. Further discussions of precision manometry are available in the literature.[1,2]

D. General Properties of Mercury. No other liquid rivals mercury as a general-purpose manometric liquid. However, mercury is not without its drawbacks: it is toxic, and it is attacked by strong oxidizing agents such as NO_2, halogens, and oxidizing fluorides. In addition, it reacts with tin tetrachloride and with a few organometallic compounds, such as dimethylcadmium. Mercury amalgamates and appreciably dissolves many metals, such as the alkali and alkaline earth metals, Zn, Cd, Sn, Pb, Bi, Ga, Au, and Tl. Traces of these metals influence the surface tension and glass-wetting tendency of mercury. Copper, Ag, Pt, and Al are somewhat less soluble, but they are soluble enough (about $10^{-3}\%$ by weight) so as to affect the surface tension. Finally, there is a group composed mostly of transition metals that are quite insoluble ($5 \times 10^{-5}\%$ or less): Cr, Co, Fe, Mo, Ni, Sb, Ti, W, and V.[3]

E. Reducing Exposure to Mercury. Elemental mercury vapor and dust of mercury compounds are absorbed through the respiratory tract, skin, and digestive tract. The current acceptable threshold level in the United States for continuous exposure to mercury in the air is 0.05 mg/m^3, which corresponds to a partial pressure of 4×10^{-6} torr. Since the vapor pressure of mercury is on the order of 2×10^{-3} torr at room temperature, it is clear that an enclosed, poorly ventilated room containing exposed mercury would vastly exceed the accepted limits. The hazards of working with mercury are minimized by adopting good housekeeping practices. The laboratory should be well ventilated, mercury vessels should be closed, and spills should be thoroughly and promptly cleaned up. After all traces of visible mercury have been picked up, the area of the spill should be treated with zinc dust or charcoal which has been treated with a solution of iodine. It is particularly undesirable to heat mercury in the open, so care must be taken to clean it from glass apparatus before glassblowing and to keep mercury off warm surfaces, such as mechanical pumps.

Mercury droplets have a great tendency to roll, making it difficult to clean up

[1]G. W. Thompson, *Techniques of Organic Chemistry*, 3rd ed., Vol. 1, Pt. 1, A. Weissberger, Ed., Interscience (New York: 1959), p. 410.

[2]A. F. Germann, *J. Phys. Chem.*, 19, 437 (1915).

[3]P. Pascal, Ed., *Nouveau Traite de Chemie Minerale*, Vol. 5, (Paris: Mason et Cie, 1962), p. 510ff.

spills. A vacuum pickup device for mercury can be constructed from a vacuum filtration flask with its sidearm connected to an aspirator, as illustrated in Fig. 7.4*a*. Another means of picking up small globules of mercury is a spiral of copper wire, as shown in Fig. 7.4*b*. A variety of mercury pickup devices is available from laboratory supply houses.

F. Cleaning Mercury. When it is received, mercury is generally covered with a scum which can foul manometers, Toepler pumps, and similar apparatus. This scum can be removed by a simple filtration procedure. A standard filter paper is folded to fit an appropriate funnel and a pinhole is pierced in the tip. The hole must be large enough so that mercury droplets will run through, but small enough so that the last drop and its associated scum are retained by its surface tension. This type of filtration process is routinely performed before filling an apparatus.

Mercury that has been contaminated by metallic impurities is cleaned by a nitric acid wash to remove most electropositive metals, and it is then distilled to

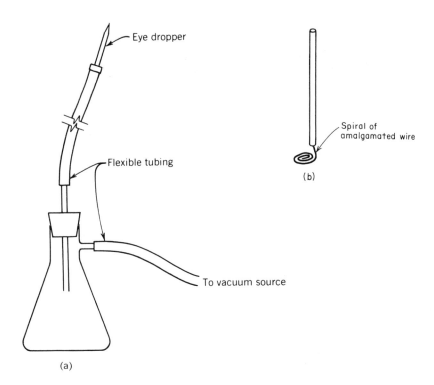

Fig. 7.4. Mercury pickup devices. (*a*) Vacuum pickup device. Collected mercury is trapped in the flask for recycling or disposal. (*b*) Amalgamated copper wire pickup device. The wire is first cleaned in nitric acid, then dipped into a solution of mercuric nitrate to give a thin coating of mercury. Droplets of mercury readily cling to the spiral and may be shaken off into a mercury waste container.

remove nonvolatile impurities. These services are available commercially and therefore will not be described further.

G. Bourdon Gauges. As an alternative to mercury manometers there is a variety of gauges based on mechanical or electrical pressure transducers. This section presents a description of purely mechanical gauges which still find use in this electronic age.[4] The metal Bourdon gauge (Fig. 7.5) is fashioned around a semicircular thin-walled metal tube with mechanical linkage to a pointer. Fused-quartz spiral gauges are also available. In this case, a thin spiral is sensitive to a pressure differential, and the deflection is balanced with air pressure in the surrounding envelope. The air pressure is then measured with a manometer.

H. Electronic Pressure Transducers. Electronic pressure transducers equipped with digital pressure readout provide a convienient and fairly compact

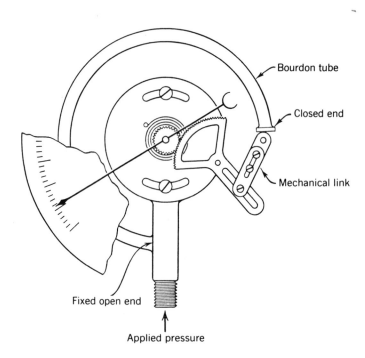

Fig. 7.5. A metal Bourdon gauge. The thin Bourdon tube is sensitive to the pressure differential between the surrounding atmosphere and the gas contained within the tube. The tube deflects an indicating needle which allows a direct pressure determination from the calibrated scale.

[4]Bourdon and spiral gauges with dial readout are available from Wallace & Tiernan, 25 Main St., Belleville, NJ 07109. Warden quartz spiral gauges are available from Ruska Instrument Corp., Box 36010, Houston, TX 77036; Texas Instruments Inc., Display Systems, Box 1444, Houston, TX 77001.

method for pressure measurement in a variety of environments. Although a wide variety of designs are available, the most common units are based on capacitance or inductance detection of the displacement of a diaphragm (Fig. 7.6).[5] The specifications on the gauge should be carefully matched to the range and accuracy which are required. Some attention must be paid to long-term drift in the gauge, especially when the gauge is exposed to changes in temperature and mechanical vibration. The portion of the gauge which is exposed to the gas can be made from a variety of materials, including fluorine-resistant metals.

7.2 MEDIUM- AND HIGH-VACUUM MEASUREMENTS (10⁻¹-10⁻⁶ TORR)

Gauges which are sensitive in this range are primarily used to determine ultimate vacuum on a system and to hunt leaks. This pressure range is measurable by a variety of gauge types ranging from the manually operated mercury-filled McLeod gauge to various electronic gauges.

A. McLeod Gauges. This gauge provides a simple and reliable way of measuring pressures greater than 10^{-6} torr. It works on the principle of compressing a large volume of low-pressure gas into a small volume where the pressure is great enough to be measured with a mercury column. Gas which is contained in the upper bulb of the gauge illustrated in Fig. 7.7 is compressed into the upper closed-end capillary, and the height differential between the mercury in this closed capillary and an adjacent open-end capillary is related to the origi-

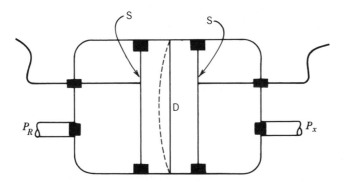

Fig. 7.6. Electronic pressure transducer. This sensing unit is based on the measurement of the capacitance between the diaphragm D and sensors S. P_x is the pressure being measured and P_R is a reference pressure which may be a high vacuum.

---

[5]Diaphragm gauges with digital readout are available from MKS Instruments Inc., 22 Third Ave., Burlington, MA 01803; Validyne Engineering Corp., 8626 Wilbur Ave., Northridge, CA 91328; Datametrics Dressler Industries Inc., 340 Fordham Rd., Wilmington, MA 01887; Fluid Precision, Inc., 110 Middlesex St., Chelmsford, MA 01863.

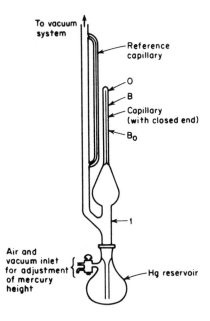

Fig. 7.7. The McLeod gauge. The principles of operation follow. Let the unknown pressure in a system be P when the Hg level is below point 1. Let the volume of the bulb and closed capillary above 1 be V, which is known. When the mercury is allowed to rise past point 1, the gas is trapped and finally compressed into the capillary. Suppose that when the mercury in the reference capillary is at 0, the mercury in the dead-ended capillary is B mm below 0 (i.e., the pressure of the compressed gas is B mm). Since the initial pressure-volume product equals the final pressure-volume product, $PV = pv$, the volume in the capillary v will be the height B times the area of the capillary bore A. Thus $P = pv/V = B^2(A/V)$. Since A and V are known and B is measured, the original pressure (P) may be calculated. Most commercial gauges are provided with a calibrated scale which presents pressures directly. Alternatively, it is possible to devise a linear scale for the McLeod gauge. In one such method the mercury height in the closed capillary is always adjusted to the same point (B_0), and then the difference in meniscus heights between the two capillaries is measured (ΔB). For this case the pressure being measured is $P = pv_0/V = (B_0 A/V)\Delta B$. As in the previous example, the quantity in parentheses represents the gauge calibration constant.

nal pressure in the vacuum system. Further details on the operation of this gauge are explained in the caption to Fig. 7.7.

A clean gauge and pure mercury are essential for reliable operation because the presence of impurities causes mercury to cling to the capillary walls. This may lead to unreliable readings and a tendency for the mercury column to break up in the capillary. If the mercury column in the capillary is broken by gas bubbles, the pressure readings are highly inaccurate. In any case, the capillary tubing is tapped to settle the mercury before a pressure measurement is obtained. It is very important to run the mercury smoothly into the upper chamber. If the mercury surges into this chamber, it may break the apparatus and shower the surrounding area with mercury. The McLeod gauges intended for the high-vacuum range are generally bulky and contain a large quantity of mercury. Because of its mass, the gauge is generally mounted on the bench of the vacuum line. The reservoir is well supported by a cork ring and/or plaster base inside a pan which is deep enough to catch the mercury in the event of a mishap.

The McLeod gauge is not suitable for the determination of pressures of easily condensed gases, such as water vapor, and it has the additional disadvantage of being slow and sometimes clumsy to operate. Because of this naturally slow response, the electronic vacuum gauges are superior for tracing leaks.

The tilting McLeod gauge (Fig. 7.8) is a simple, inexpensive, and portable gauge which may be used to measure pressures down to about 10^{-3} torr. These gauges are very useful for checking rough vacuum systems, Schlenk systems, and for the calibration of thermal conductivity vacuum gauges.

B. Thermal Conductivity Gauges. Included in this category is a series of similar gauges which go under a variety of names: Pirani, thermocouple, and thermistor gauges. These simple and relatively inexpensive gauges contain a heated element, and the temperature of that element is detected by means of a thermocouple or thermistor (Fig. 7.9). Direct readout in pressure is provided by means of a calibrated dial. It is common for these units to be calibrated over the 1–10^{-3}-torr range; however, at either end of the scale the readings are approximate. A gauge of this type is useful for measuring the pressure in the antechamber of a glove box, for monitoring the fore pump pressure in large vacuum sys-

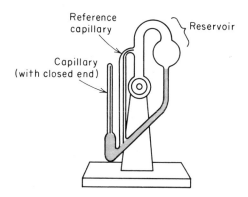

Fig. 7.8. Tilting McLeod gauge. The gauge rotates around the center section. This center section also contains a nipple (extending out from the back of the drawing) which may be attached to the vacuum system through a hose. For a more permanent arrangement, the center section is attached to a vacuum line by means of a standard taper joint and the base is discarded. Initially, the gauge is rotated so that the mercury runs into the reservoir. When pressure equilibrium is established with the system being measured, the gauge is rotated back so that gas is compressed into the dead-end capillary and the top of the mercury in the reference capillary coincides with the top end of the dead-end capillary. Generally, the gauge is provided with a scale which may be read directly in terms of pressure. These gauges cover the medium-vacuum range, but generally do not extend into the high-vacuum region.

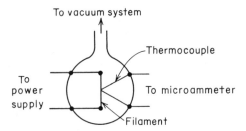

Fig. 7.9. Sensing element for the thermocouple vacuum gauge. The thermocouple is in contact with a heated filament and measures its temperature. A variant involves a thermistor which serves both as a heating and a sensing element.

tems, and for chemical vacuum systems in which high vacuum is not required. The thermal conductivity of a gas is dependent not only on the pressure, but also on the heat capacity, molecular weight, and accommodation coefficient (efficiency of energy transfer with a surface) of the gas being measured. Furthermore, the accommodation coefficient will vary if the surface is coated with foreign matter. Therefore, these gauges are generally not suitable for accurate pressure measurement, but this is of little consequence in their most frequent application as sensors of the general degree of vacuum in a system.

One of the most troublesome aspects of the thermal conductivity gauges is the altered response caused by the pyrolysis of thermally unstable molecules on the hot sensing element. To minimize this problem, the gauge should be turned off, or the sensing tube isolated, when thermally sensitive compounds are handled. The calibration of the gauge can be checked periodically with a tilting McLeod gauge and the sensing tube replaced when the unit gets badly out of calibration. The sensing head is generally constructed from metal with a tube inlet. This sensor may be attached to a glass vacuum system by the use of vacuum wax to secure the inlet tube inside a glass tube on the vacuum system. To facilitate making a good seal, the latter tube should have an inside diameter just slightly larger than the outside diameter of the tube on the sensor. Often the tube on the sensor has pipe threads and this may be attached to a pipe thread-to-swage union. As described in Chapter 8, a swage joint fitted with a Teflon ferrule can be attached to glass tubing of appropriate diameter.

C. Ionization Gauges. This rather broad class of gauges includes cold-cathode ionization gauges (sometimes called Penning gauges) and thermionic or hot-cathode vacuum gauges (one highly efficient version is called the Bayard–Alpert gauge). These gauges are mainly used in the pressure range below 10^{-3} torr, and the thermionic gauges extend into the ultrahigh vacuum range (10^{-12} torr). The operation of these gauges is based on a common principle: the measurement of the positive ion current from gas molecules which have been ionized by electron impact.

In the cold-cathode gauge, a glow discharge is initiated by a very-high-voltage starting pulse and is maintained by an applied high voltage. A schematic of the sensing head is shown in Fig. 7.10. A magnetic field is employed in this sensor to confine the electrons to a spiral path and thus increase the probability of collision with gas molecules. The sensitivity to various gases varies with the nature of the gas, with the sensitivity for helium and hydrogen being low and that for mercury and polyatomic molecules such as acetone being high. In its most common application, judging the initial degree of evacuation of a system, the factor of 30 or so in sensitivity to various common gases is of little importance.

The cold-cathode gauge is a good choice for measuring the initial vacuum in chemical high-vacuum systems because the pressure range is appropriate and the gauge is robust. Standard commercial units are available for measuring pressures in the approximate range of 10^{-2} to 10^{-7} torr, and the range of most interest for synthetic chemical systems extends from this upper limit to about 10^{-6} torr. The sensing element collects debris from the fragmentaion of polyatomic molecules, and therefore it is necessary to turn off the instrument or isolate the sensing element when significant pressures of polyatomic molecules are present. Fortunately, the gauge is tolerant to a fair amount of deposit and the sensing head can be cleaned by scraping out or dissolving the deposits. The re-

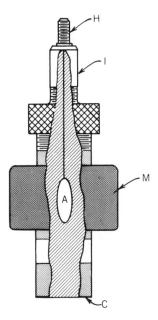

Fig. 7.10. Cutaway of a cold cathode (Penning) gauge. The wire anode (A) is connected through a ceramic insulator (I) to the high-voltage lead (H). Under the influence of the magnetic field created by M, the electrons travel a long spiral path between the body and the anode. The current, which arises from electrons and positive ions of the ionized gas molecules, is related to the pressure. The open end (C) is attached to the vacuum system.

sponse of the ionization gauge is fast, which makes it an excellent tool for tracing leaks. The sensing element for a cold-cathode gauge is generally contained in either a glass or metal envelope. If a unit with a glass envelope is used, it may be sealed directly to the vacuum system and cut off from the system when the time comes for cleaning. Metal heads may be waxed to a large tubular opening on the vacuum system, or attached by an O-ring joint, such as a Cajon Ultra-Torr union (see Section 8.1.B).

In the cold-cathode gauge a steady-state concentration of electrons is maintained by electron impact on the molecules being measured. By contrast, the primary source of electrons in hot-cathode gauges, such as the Bayard-Alpert gauge, is a hot filament. These electrons are accelerated toward a positive grid, and in this region molecules are ionized by electron impact. The positive grid collects the electrons and permits positive ions to pass through to an electrode which is biased negative with respect to the filament. This positive ion current is then related to pressure. These types of gauges have excellent high-vacuum sensitivity and are routinely used in high-vacuum and ultrahigh-vacuum systems. For chemical systems where there may be a moderately high background of large molecules, the hot-filament gauges are fairly readily fouled, and if exposed to air the hot filament is burned out. For these reasons, the hot-cathode gauge is preferred for clean, very-high-vacuum systems and is rarely used for vacuum lines primarily devoted to syntheses.

Both the thermionic and cold-cathode gauges exert a significant pumping action on a vacuum system due to the breakdown and deposition of molecular ions created by electron impact. These gauges are prone to degas deposited material when turned off, or soon after they are turned on.

D. Residual Gas Analyzers.   A residual gas analyzer is basically a small mass spectrometer which provides a method of identifying the foreign gas in a vacuum chamber as well as judging its pressure. These devices are widely used on ultrahigh-vacuum apparatus to judge the nature of the background gas and to detect the source of contamination. They also are used to diagnose the performance of a variety of apparatus employed in modern thin-film research and technology, such as vacuum evaporators, plasma etching devices, and the like. The residual gas analyzer is indispensable in the ultrahigh-vacuum apparatus employed for the study of chemistry on single crystal surfaces, where the nature of contaminant gases is important. This type of apparatus is also useful for studying the temperature-dependent desorption of molecules from surfaces, and to identify gaseous products in chemical and catalytic systems.

Typical residual gas analyzers employ electron impact to generate ions and a quadrupole "mass filter" to obtain the mass-to-charge ratio of these species. Mass ranges for these analyzers vary: 2–80 atomic mass units for small units; 1–500 on larger, more expensive gas analyzers. Electron impact ionization of the gas is achieved using a hot filament as the source. Many instruments are equipped with a second filament which can be switched into use in the event of a failure of the first. When a robust detector of moderate sensitivity is needed, a

Faraday cup is used to collect ions and thus provide a direct measure of the ion current. For higher sensitivity and rapid response, an electron multiplier detector is employed. In this device the accelerated ions impinge on a surface with the emission of secondary electrons. These are multiplied by a cascade along either a channel or set of plates having a large potential gradient.

The residual gas analyzer sensing head is generally supplied on an ultrahigh-vacuum flange which can be readily accommodated on most metal vacuum apparatus.

7.3 LEAK DETECTION

A. Strategies. Leaks are a ubiquitous problem with vacuum systems. They come in two varieties: real and virtual. A real leak involves an opening in the vacuum chamber. Examples are a pinhole or hairline crack in the glass, or a stopcock in which the grease has channeled. Virtual leaks, on the other hand, are contained within the vacuum system and might include such things as the outgassing of stopcock grease or O-rings which have absorbed solvent vapors. The walls of a glass apparatus present a large virtual leak as the moisture is desorbed from the freshly evacuated apparatus. A knowledge of the prior use of the apparatus will often permit one to spot a virtual leak. Alternatively, the outgassing of condensable materials can be verified by cooling a section of the vacuum system with liquid nitrogen while the pressure is checked with a vacuum gauge. Another method of differentiating real and virtual leaks is to pump out the section to be tested, isolate the section, and measure and record the pressure as a function of time. The rate of pressure rise for a real leak is constant, whereas the virtual leak will tend to level off as the effective vapor pressure of the outgassing material is reached. The following sections will be devoted primarily to the detection of real leaks.

It is important to conduct the search for leaks in a thorough and methodical fashion, as outlined in Table 7.1. The examples given here are for a glass system with greased stopcocks and joints, but the reader should have little difficulty in applying the same strategy to other types of apparatus. In many instances there may be a suspect site which a quick check will reveal to be the source of the leak. If, however, there is no obvious source, a methodical check should be inititated. For any system, the vacuum pump should be checked first to ensure it is delivering the appropriate vacuum. In a typical preparative vacuum system, stopcocks, valves, and joints are the most common sources of leaks. Freshly greased joints, particularly those which have been greased too heavily, are likely to develop leakage paths which generally take the form of dendritic channels. Old greased joints or those which have been subjected to solvent vapors may develop leaks when turned, because the grease film is too thin. These leakage paths are generally evident in the form of fine striations which extend around the diameter of the stopcock but interconnect at various points. Thus a visual inspection of the greased joints and stopcocks may turn up the source of the leak. Once a suspect

Table 7.1. Summary of the Strategy for Locating and Controlling Leaks

Preliminary Survey of Valves

Preliminary visual check of any greased stopcocks and valves. (Look for striations or channels in the grease.)

Methodical Search for Leak

1. Check pumping system to be sure it is achieving a good vacuum. If necessary, correct problem with fore pump, diffusion pump, or cold trap.
2. Pump out entire system and turn off stopcocks or valves to isolate every system possible.
3. After a short wait, open section to the vacuum gauge and note whether or not there is a large pressure rise.
4. When a suspect section is located, pump it out, isolate it from the pump and measure the pressure rise at regular time increments. A steady pressure rise indicates a real leak. A leveling off of pressure rise with time indicates a virtual leak.
5. Inspect greased stopcocks and joints in the vicinity of the leak, and turn the stopcock or joint while noting the pressure. Run a Tesla coil over glass-blown joints to locate pinholes. Inspect O-ring contact on glassware. Squirt CH_3OH on suspected parts while noting pressure on an ionization gauge. Use a halogen or helium leak tester (a necessity on metal systems).
6. Clean system and/or bake it out if virtual leaks are indicated.

Fixing a Leak

1. Regrease faulty stopcocks or joints.
2. Clean O-rings and/or replace O-rings on leaky O-joints or valves.
3. Tighten Swagelok joints or remake the joint with new tubing and ferrules.
4. Troubleshoot the seat, packing, or bellows on metal valves.
5. Repair pinholes in glass or poorly welded or brazed metal parts.

joint or stopcock is found, its potential role as a leakage site can be checked by rotating it slightly while checking an electronic vacuum gauge for fluctuations in the vacuum. Often O-ring joints or glass valves containing O-ring seals are potential sites of leakage, but here the problem is more difficult to spot visually.

If visual inspection of the most suspect joints or stopcocks fails to reveal the leak, a systematic isolation of parts of the vacuum system is in order. For a vacuum line of conventional design (e.g., a line approximating that in Fig. 5.2), it is generally best to turn off all stopcocks which interconnect the various parts of the vacuum system. As a result, the high-vacuum manifold is isolated from the pumps and the rest of the line, and it is checked by determining if there is a steady pressure rise in that section. If this section appears to be intact, but it does not pump down well when reopened to the pumps, the vacuum source is suspect. In the event that the vacuum source and main manifold are not at fault, each section of the vacuum system is then opened to the high-vacuum manifold, with an eye on the pressure gauge to indicate the leaking section. Once the general region of the leak is identified, this part of the apparatus is thoroughly checked to pinpoint the leak. A visual check of joints and stopcocks may again be in order, or the search may be conducted with one of the devices mentioned below.

B. Leak Detection Using a Tesla Coil. Pinholes are a major source of leaks, particularly in newly constructed glassware. One of the common devices for locating these leaks is the Tesla coil, which provides a high-voltage, high-frequency spark discharge in air. The probe of the Tesla coil is touched to the glass apparatus at various spots until a glow discharge is initiated inside the vacuum chamber. The probe is then passed over the suspect parts of the glassware, and the presence of a pinhole is revealed by an intense blue spark which jumps through the hole. The Tesla coil should never be used around thin glass, such as bread-seals, since it will puncture holes in the apparatus. Similarly, it may puncture O-ring materials and generally is not useful around greased joints. The Tesla coil is also ineffective around metal parts, which simply serve as a ground.

The color of the glow discharge inside the apparatus is characteristic of the gases present and of some help in locating the leak. For example, if the atmosphere in the vacuum system is primarily mercury, a light blue glow is seen; but if there is an air leak, the discharge takes on a more purple color. When the pressure in the system drops below approximately 10^{-3} torr, the glow discharge cannot be sustained, and this is sometimes useful as a crude indication of the degree of vacuum.

C. Leak Detection Using Vacuum Gauges. Reference has already been made to the use of an electronic vacuum gauge in hunting leaks. A cold-cathode gauge is particularly useful because it has rapid response and operates well in the presence of small leaks. Because of this rapid response, acetone or methanol can be squirted on the suspected area. If a hole is present, a momentary pressure drop will be observed because the acetone or methanol molecules are slower than air to diffuse through the hole. Sometimes a helium stream is directed on the suspected leak; in this case a pressure rise is noted when the helium passes over the site of the leak, because helium diffuses more rapidly than air. It also is sometimes possible to use a puttylike material, such as Apiezon Q, to cover temporarily a suspected leak.

If the leak path is tortuous, these techniques do not work well and a leak detector may be more suitable. For example, leaks in swage-type fittings (Section 10.3.C) and valves of various types may be difficult to detect using these methods.

D. Leak Detection Using Helium or Halogen Leak Detectors. Commercial leak testers have high sensitivity and are very useful with metal vacuum systems. Only two common types, one based on halogenated compounds and one based on helium, will be covered here.

The halogen leak tester is widely used in refrigeration applications, but it is sufficiently sensitive for many types of chemical vacuum systems. The apparatus is filled with a halogenated hydrocarbon and the probe of the tester is passed over the suspected area. The gas that is collected by the probe is passed over a hot platinum wire with the resulting formation of positive ions which are collected and measured with a microammeter.

Helium leak testers have far higher sensitivity than those based on halogens. In this case, helium is detected by a small mass spectrometer set on mass 4. In the least sensitive mode the apparatus is filled with helium and the suspect sites are "sniffed" with a probe connected to the vacuum chamber of the mass detector. In a more sensitive mode the leak detector is connected to the vacuum system often as the sole pump on the system. A stream of helium is then passed over the suspect sites for the leak. An audible and/or meter signal is used to indicate the amount of helium reaching the mass detector. Since the natural background at mass 4 is very low, these testers can be highly sensitive and very specific in pinpointing leaks. The helium leak tester is, however, fairly complex and expensive, since it includes, in addition to the mass spectrometer, a vacuum system complete with liquid nitrogen trap, diffusion pump, fore pump, and often a large roughing pump.

E. Leak Detection Using Residual Gas Analyzers. The residual gas analyzer, which has been described in Section 7.2.D, is an excellent device for locating both virtual and real leaks. Most of these instruments can be scanned through the low molecular weight mass range. This mass spectrum allows ready discrimination between virtual leaks from solvents and the like, versus real leaks, which show up with high peaks corresponding to the atmospheric gases. The location of the specific site of leakage may be accomplished by setting the instrument on mass 4 and passing a small jet of helium over the suspected areas.

7.4 FLOW MEASUREMENT

When working with apparatus such as catalytic flow reactors, the measurement and control of gas flow becomes as important as the measurement of pressure just described for vacuum line work. Some of the more common methods for measuring and controlling gas flow in gas-handling systems will now be discussed.

Gas flow is usually quoted in terms of velocity (cm/s) or "mass flow" (standard cm³/s). Velocity may be measured by a venturi tube pointing into the flowing stream with a pressure transducer or manometer connected to the venturi tube. The two schemes which are more attractive for small-scale apparatus are the rotameter, which is based on a "float" which is bouyed by the stream of gas, and a variety of thermal methods based on thermal conductivity or heat capacity. The thermal methods are readily adapted to digital output and automatic flow control. Although the temperature of the operating sensor element is not high (generally under 100°C), thermally sensitive compounds can be pyrolyzed in this type of flow meter.

A. Flow Measurement by Displacement. A variety of flow meter designs are based on the positive displacement of a small amount of easily visualized material which does not alter the flow rate. A simple and easily constructed mass flow meter of this type, the soap-film meter, is based on timing the displacement of a soap film up a buret tube (Fig. 7.11). This type of flow meter is often placed at the exit of a gas chromatograph or small flow reactor. It is very

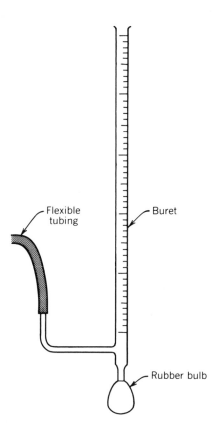

Fig. 7.11. Soap-film flow meter. Soap solution is poured into the rubber bulb so that the solution level is just below the level of the sidearm. After attaching the flexible tubing to the flowing gas source, the rubber bulb is gently squeezed. This allows the flowing gas to percolate through the soap solution, picking up a thin film of soap which will travel up the buret. By noting how long it takes for the film to travel between two points on the buret, the flow may be determined.

useful for occasional measurements, but cumbersome if the flow must be monitored frequently.

B. Flow Measurement by Orifice Area.   The simplest and most popular laboratory mass flow gauge is the rotameter, which consists of a float in a vertical tapered tube with the larger diameter at the top and the gas inlet at the bottom (Fig. 7.12). The float maintains a constant pressure differential on the gas flowing past it, and as the float rides higher a larger annular opening is available for the gas. Graduations on the side of the glass or clear plastic tapered tube are thus related to the flow rate. These graduations are calibrated for a specific gas (often dry air); a calibration chart is needed for other gases. The calibration of a rotameter is dependent on the pressure in the system. If true "mass flow" (molar flow) is required, a pressure correction is necessary.

C. Flow Measurement by Pressure Drop across an Orifice.   Another common scheme for the measurement of flow is based on the determination of the pressure drop on either side of a constriction, such as an orifice or venturi. Either a liquid-filled differential manometer or a pressure transducer with associated digital readout may be used for this pressure measurement. The flow rates determined by these meters are in units such as cm^3/s, and it is necessary to make a correction for total pressure to convert these to standard cm^3/s or mol/s.

A popular series of mass flow meters is available which is based on the use of a thermal sensor in a side stream (Fig. 7.13). Generally, these flow meters are

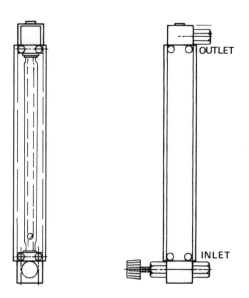

Fig. 7.12.   Rotameter flow meter. The ball is bouyed by gas flow up the conical tube. A graduated scale is provided for visual measurement of the flow rate.

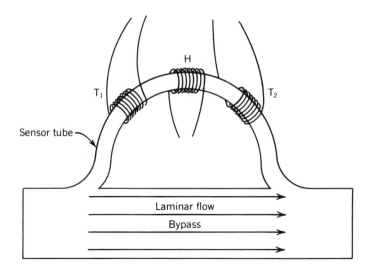

Fig. 7.13. Schematic of a mass flow meter. The flow is proportional to $(T_2 - T_1)^{-1}$. This type of meter is often employed in an electrically controlled flow regulator. T_1 and T_2 are temperature sensors; H is a heater.

supplied with a digital readout calibrated for the mass flow of one specific gas. Thermally sensitive compounds may be pyrolyzed on the thermal element, and the small sensor channel can be blocked by the decomposition products. The calibration on these meters is dependent on the heat capacity of the gas being handled and, to a somewhat lesser extent, on the viscosity of the gas. The calibration factor can be electronically altered for convenience in handling different gases.[6]

7.5 FLOW CONTROL

A. Flow Control without Feedback. Flow can be controlled by means of a needle valve if the pressure drop across the valve is constant. The pressure on the upstream side often can be held constant with a single- or two-stage mechanical diaphragm regulator (Section 10.1.B). If the stream of gas does not experience a variable constriction after the needle valve, the above combination provides a simple and convenient means of providing a steady flow. Often an arrangement such as this is used in conjunction with a rotameter or electronic mass flow meter (Fig. 7.14).

[6]Flow meters based on thermal sensors and electronic flow controllers are available from the following companies: MKS and Datametrics, listed in footnote 5; Brooks Instruments Div., Emerson Electric Co., 407 W. Vine St., Hatfield, PA 19440; Teledyne Hastings-Raydist, Hampton, VA 23661; Tylan Inc., 23301 S. Wilmington Ave., Carson, CA 90745; Unit Instruments Inc., 1140 E. Valencia Dr., Fullerton, CA 92631.

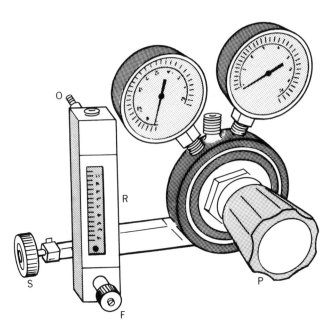

Fig. 7.14. Combination of two-stage pressure regulator (P), shutoff valve (S), flow control valve (F), and rotameter (R). O is the gas outlet.

D. Flow Control with Feedback. More positive feedback control of the flow rate is needed for applications such as flow reactors for the epitaxial deposition of semiconductor films, and catalytic flow reactors. Several electronic flow controllers are available. One of the most commonly used versions is based on the mass flow meter with a thermal sensing element, mentioned above, and feedback control to a valve, which typically is either actuated thermally or magnetically.[6] Some of these valves have rather small orifices which can be clogged by the decomposition of reactive molecules, for example SiH_4 and $Ga(CH_3)_3$, which may react with traces of oxygen. As with the flow meters based on the same principle, these controllers regulate the molar flow, independent of total pressure. A calibration factor is necessary for different gases, but this can be determined from the ratio of heat capacities for gases of similar nature. A large change in the character of the the gas—say, hydrogen in place of nitrogen—requires experimental recalibration.

GENERAL REFERENCES

Benedict, R. P., 1984, *Fundamentals of Temperature, Pressure, and Flow Measurements*, 3rd ed., Wiley, New York. Part II of this book, entitled "Pressure and Its Measurement," presents an excellent overview of the concept and standards of pressure, and the use of mechanical and electronic transducers to measure pressure. "Flow and Its Measurement" is reviewed in Part III, including chapters on the concept of flow rate and theoretical rates for constant-density and compressible fluids in closed-channel flow.

Introduction to Helium Mass Spectrometer Leak Detection, 1980, Varian Associates, Palo Alto. This reference presents a good overview of the fundamentals and methods of leak detection, followed by a detailed discussion of the helium leak detector.

Leak Detector Manual, 1981, General Electric Co., Lynn, Mass. The manual focuses on halogen leak detector methods developed by this company.

O'Hanlon, J. F., 1980, *A Users Guide to Vacuum Technology*, Wiley, New York. This book is a good source of information on vacuum gauges and residual gas analyzers.

Thompson, G. W., 1959, *Techniques of Organic Chemistry*, Vol. 1, Part 1, 3rd ed., A. Weissberger, Ed., Interscience, New York.

8

JOINTS, STOPCOCKS, AND VALVES

This chapter describes the various stopcocks, valves, and joints used in glass vacuum and inert-atmosphere systems. Although greased stopcocks and joints are familiar and straightforward in design, their use on high-vacuum systems requires more than the usual amount of attention to detail. The newer O-ring-sealed joints and valves, which are taking the place of greased stopcocks and joints in many applications, are also discussed here. Metal bellows valves are more leak-tight than glass valves or stopcocks, but their use on glass apparatus requires special designs which are outlined at the end of this chapter.

8.1 JOINTS

A. Standard Taper and Ball Joints. These joints (Fig. 8.1) have been used for many years on vacuum and inert-atmosphere systems, and the availability of a variety of synthetic greases and waxes has extended their utility. Of the two, the ball-and-socket joint permits more flexibility in the orientation of the two halves. When this flexibility is important, this type of joint is obviously to be preferred, even though the ball joint is somewhat more prone to leak than a standard taper joint. Both metal ball joints and metal standard taper joints are available,[1] and these items provide one means of joining metal and glass apparatus.

---

[1]Metal ball joints, metal standard taper joints, and O-ring joints with a step tooled into the groove are available from Kontes Glass Co., P.O. Box 729, Vineland, NJ 08360. The first two items are also available from Ace Glass Co., P.O. Box 688, Vineland, NJ 08360.

152

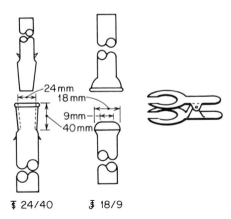

Fig. 8.1. Standard taper (⊤) and spherical joint (⊕). When the joints are lubricated with grease, they must generally be held together. Springs or rubber bands are frequently employed on standard taper joints, while a spring-loaded clamp (illustrated here) or a screw clamp (illustrated in Fig. 8.3) is used with ball joints. The method used for specifying joint sizes in the United States is illustrated, and it is described in detail in National Bureau of Standards, Commercial Standard CS 21-39.

Teflon sleeves are available which may be used to replace grease as a sealing agent in standard taper joints. These sleeves are useful on still pots and in other harsh but not highly demanding situations; however, it has been the authors' experience that they are not suitable for high-vacuum applications or in situations requiring very good exclusion of the atmosphere. To achieve good vacuum-tight performance with standard taper joints, the grease may be applied in several thin stripes which run parallel to the direction of the male joint (Fig. 8.2). The joint is then worked together with an oscillating motion which squeezes out the pockets of air and leads to a thin, uniform film. Good-quality stopcock grease containing a minimum of volatiles and uncontaminated by dirt or dust is essential, because volatiles outgas badly and particles interrupt the film and produce leaks. The film of grease must be thin because the pressure differential across the joint will pull channels in a thick film. Silicone and hydrocarbon greases are the most popular; the chemical and physical characteristics of these and other common greases are given in Table 8.1.

In general, stopcocks which have been disassembled should be cleaned thoroughly before they are regreased and reassembled. Hydrocarbon solvents will remove most greases. When the glassware is being cleaned rigorously, hydrocarbon residues can be removed in an acid-dichromate cleaning bath. Residues from silicone stopcock grease are much more tenacious. Thin films of such residues can be removed by means of alcoholic KOH (methanol nearly saturated with KOH). Ground-glass items, and particularly fritted items, should not be left in the alcoholic KOH bath for extended periods, since this cleaning solution slowly etches glass.

When a joint is only occasionally opened and flexibility at the joint is not

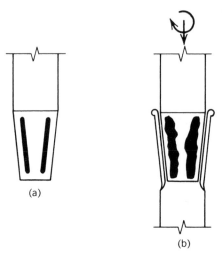

Fig. 8.2. Application of grease to joints and stopcocks. (*a*) Grease is applied in several thin stripes which run parallel to the direction of the male joint. A wooden splint may be used to apply the grease to avoid contamination by moisture or dust on fingers. (*b*) The joint is worked together with an oscillating motion and slight steady pressure.

needed, it is generally preferable to use wax instead of grease. Wax makes a much more permanent joint than grease because it does not flow out of the joint and is somewhat less sensitive to solvent vapors. Two excellent waxes are Apiezon W, a black hydrocarbon wax, and Kel-F 200, a colorless chlorofluoro-carbon of low vapor pressure and good resistance to chemical attack. The characteristics of these waxes are given in Table 8.2. Both halves of the joint are heated to somewhat above the melting point of the wax using a small gas burner flame or heat gun. A stick of the wax may then be rubbed on the male joint, preferably in stripes as described above. The two warm parts are joined and the wax is worked into a uniform film by oscillating the joints and pushing the two parts together, usually with additional heating of the outer joint. Temperatures well above the melting point cause breakdown of the wax; this is particularly true of DeKhotinsky-type waxes, which polymerize when heated too strongly. To open a waxed joint, it is heated uniformly and gently to just above the melting point and disassembled. It is most satisfactory to clean off old wax with the appropriate solvent and apply fresh material before the joint is reassembled.

B.   O-Ring Joints.   The major difficulty with greased and waxed joints is their failure to hold vacuum when exposed to solvents, heat, and some chemicals. For many applications O-joints overcome these limitations. A successful glass O-ring joint which is in wide use is illustrated in Fig. 8.3, where it may be seen that both halves of the joint are identical. They mate with an O-ring fitted into a groove, and the two parts are held in place with a clamp. The two halves of these

Table 8.1. Some Common Greases for Vacuum Apparatus

Brand and Type	Approximate Vapor Pressure (mm, at room temp.)	Application	Approximate Usable Range (°C)	Resistance to Organic Solvent Vapors	Chemically Attacked by
Hydrocarbons					
Apiezon L	10^{-10}	Ground joints	Max. 30	Poor	Reactive halides such as BCl_3, and very strong oxidizing agents such as O_3
Apiezon M	10^{-7}	Ground joints	Max. 30	Poor	
Apiezon N	10^{-8}	Stopcocks and joints	Max. 30	Poor	
Apiezon T	10^{-8}	Stopcocks and joints	Max. 110	Poor	
Halocarbon					
25–55	$< 10^{-3a}$	Stopcocks and joints	Max. ca. 30	Poor	Strong reducing agents such as alkali metals, and strong nucleophiles such as alkyl phosphines
Silicone					
Dow Corning Hy Vac	$< 10^{-6}$	Stopcocks and joints	Ca. -20 to $>$ 100	Fair	Tends to cake after long exposure to NH_3 gas Reactive metalloid fluorides like BF_3

[a]Many samples of this grease contains some low-molecular-weight volatiles and silica gel. Halocarbon Products, 82 Burlews Court, Hackensack, NJ 07601.

Table 8.2. Some Common Hard Waxes for Vacuum Apparatus

Brand and Type	Approximate Maximum Usable Temp. (°C)	Temperature for Application (°C)	Resistance to Organic Solvents
Apiezon W[a] (hydrocarbon)	80	100	Hydrocarbons and nonpolar solvents—poor Lower alcohols—good
Sealstix[b] (improved DeKhotinsky)	100	140	Hydrocarbons—good Alcohols, acetone, ethyl acetate, and dioxane—poor
Kel-F 200[c]	40	90	Hydrocarbons and nonpolar solvents—poor Lower alcohols—good

[a]Available from most scientific suppliers.
[b]Central Scientific Co., Chicago, IL, and many other scientific suppliers.
[c]Minnesota Mining and Manufacturing Co., St. Paul, MN.

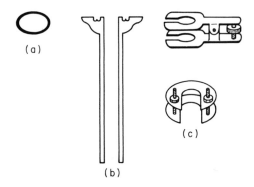

Fig. 8.3. Cross-section of a Urry-type glass O-ring joint and two types of screw clamps. (*a*) An O-ring. (*b*) Cross section of a Urry-type O-joint. Note the ridge which is tooled into the groove. (*c*) Two types of joint clamps. The upper one is manufactured by A. H. Thomas Co., Philadelphia, PA 19105.

O-ring joints are identical, which gives flexibility in apparatus design and use. Although these joints are nominally standardized so those made by the various manufacturers will take the same size O-rings and clamps, there are important differences in the way the O-ring groove is tooled. Some commercially available O-ring joints have a groove for the O-ring which has a simple rounded cross section. This type of joint is much more prone to leak than joints made with a ridge or step tooled into the the the O-ring groove.[1] The step or ridge provides higher pressure on the O-ring and thus gives a better seal than the designs in which the pressure is distributed more uniformly over the groove. Since O-rings are manufactured from elastomeric materials, they are subject to the absorption of solvents with attendant swelling. In addition to the undesirable changes in dimensions this causes, swelling also leads to a loss in strength and, in extreme cases, disintegration of the O-ring may result. The problem is considerably less serious than for greases and waxes, and it is often possible to choose an O-ring material which is only slightly affected by the solvents or chemicals being handled. To aid in this selection a summary of elastomer properties is given in Table 8.3, and a more extensive discussion of elastomers is given in Appendix III. The bulk of the laboratory applications is served by two types of O-ring materials: hydrocarbon rubbers, such as ethylene-propylene rubber or butyl rubber, and halocarbon rubber, such as Viton. The former withstands polar solvents and strong reducing agents, whereas the latter holds up to nonpolar solvents, oxidizing agents, and Lewis acids. Teflon O-rings and Teflon-coated O-rings are available, but most of these are not sufficiently compliant to provide reliable leak-tight seals.

The Solv-Seal is a variant on the above O-joint design which reduces contact of the O-ring with solvents.[2] As illustrated in Fig. 8.4, this joint consists of a pair of O-rings fitted to the outside of a Teflon cylinder and a pair of glass joints which mate with this seal. The cylinder fits inside the glass parts and the parts are clamped together with a standard compression clamp of the type illustrated in Fig. 8.3. The vacuum seal is made at the concentric O-rings, but the supporting Teflon cylinder also mates with the glass parts sufficiently well to reduce contact between the O-rings and solvents. As with the standard O-ring joint described above, the symmetry of the two halves of the Solv-Seal joint and the ability to positively clamp the two halves together make these joints especially versatile. Both types of O-ring joint are very useful in the Schlenk-type inert-atmosphere glassware (see Chapter 1), and on high-vacuum systems in which organic solvents are handled extensively.

There is a variety of O-ring fittings which attach to straight glass tubes (Fig. 8.5). The outer body of these fittings is constructed from metal or plastic and contains a threaded nut which either directly or indirectly compresses an O-ring between the outer wall of a tube and the inner wall of the fitting.[3] These fittings provide good vacuum-tight performance and are especially useful for attaching NMR tubes and storage tubes to a vacuum system so they may be sealed off. They are also convenient for joining metal and glass tubing and find application in situations where the two parts of an apparatus must rotate while a leak-tight seal is maintained (for example, the vacuum line filtration apparatus described

[2]Solv-seals are available from Fischer and Porter Co., Lab-Crest Div., County Line Road, Warminsiter, PA 18974.

[3]The concentric O-ring fittings are available from Crawford Fitting Co., 29500 Solon Rd., Solon, OH 44139 (under the trade name of Cajon Ultra-Torr fittings); and Kontes Glass Co. (address in footnote 1; items K-179900 and K-179910).

Table 8.3. Properties of some Elastomers[a, b]

Elastomers	Solvent Sensitivity				Chemical Reactivity				Approximate[c] Usable Temp. Range (°C)
	Hydrocarbons	Polar Organics	Alcohols	Chlorinated Hydrocarbons	Strong Oxidants	Strong Reductants	Covalent Halides	Conc. Sulfuric	
Butyl Rubber	P,P	F,G,P,F,F	G,F	P	P,F	—	G,P	F	−54, 150°
Buna N	G,P	P,P,P,P	G,G	F to P	P,P	—	G,P	P	−54, 120°
EPR	P,P	F,G,P,F,F	G,F	P	P,F	G	G,P	F	−54, 150°
Chloroprene	F,P	P,F,P,P	G,F	P	P,P	—	P,P	P	−54, 150°
Silicone	P,P	P,P,F,P	F,F	P	P,F	—	F,−	P	−80, 230°
Viton®	G,G	P,P,P,P	G,G	G	G,G	P	G,G	G	−50, 300°
Solvent code[d]	1,2	3,4,5,6,7	8,9		10,11	12	13,14		

[a]G = Good, F = Fair, P = Poor. All ratings are for room temperature.

[b]Adapted in part from data supplied by E. I. duPont and by Parker Seal Co.

[c]For the dry O-ring in air. The low-temperature limits are only attainable with special formulations.

[d]1, n-hexane; 2, benzene; 3, pyridine; 4, acetone; 5, diethyl ether; 6, ethyl acetate; 7, nitromethane; 8, methyl alcohol; 9, n-hexyl alcohol; 10, bromine; 11, 50% chromic acid; 12, Na in liq. NH_3; 13, stannic chloride; 14, sulfur chloride.

Fig. 8.4. Solv-seal joint. A pair of O-rings form the seal between the Teflon cylinder and glass joints. The cylinder fits inside the glass parts. A compression clamp (see Fig. 8.3c) holds the joint together.

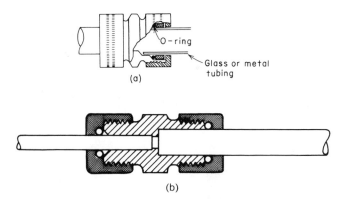

Fig. 8.5. A cutaway illustration of O-ring coupling devices. (a.) The body and knurled nuts are made of metal (Cajon union). (b). Body and compression nuts made of plastic (Kontes).

in Chapter 9). Fittings such as the Cajon Ultra-Torr require tubing of closely matching diameter.

Still another O-ring joint design is based on a glass fitting with an inner thread[4] (Fig 8.6).

C. Swage-Type Fittings.
These fittings are widely used for joining metal tubing, but they can also be used with glass tubing. Swage joints are described in detail in Section 10.3.C, which should be consulted for details. Although these fittings are used successfully for glass-to-glass seals, the aforementioned O-ring joints are generally more satisfactory because they are easier to tighten and less likely to lead to stress on the glass when they are installed. The major application of swage-type fittings is in joining metal and glass parts: for example, to attach a metal gas delivery line to a glass apparatus, or to attach metal valves to a glass system. A tapered Teflon ferrule is used to form the seal between the fitting and the glass tube. Since the compliance of this ferrule is not great, it is essential to use a ferrule of the correct dimensions to match the glass tubing. This may require the use of medium-walled tubing, which is sized in English units, to mate the standard swage joints, which are generally supplied with dimensions in English units. It is also possible to obtain fittings and ferrules made to metric dimensions[5] which will mate directly with the appropriate-size stan-

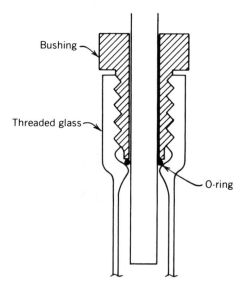

Fig. 8.6. Ace threaded connector. This connector provides an O-ring seal between a threaded glass tube and an inner tube which may be glass, plastic, or metal.

---

[4]Threaded O-ring connectors are available from Ace Glass Co. (address in footnote 1).

[5]Swage-type fittings to metric and English dimensions are available from Crawford Fitting Co. under the trade name Swagelok (address in footnote 3).

dard walled glass tubing, which is also made to metric dimensions. To avoid cutting the ferrule or cracking the glass tube inside the swage fitting, it is best to finish the glass tube with a clean cut perpendicular to the tube axis, and to remove the sharp edge with very light sanding or a very light fire polish. Because Teflon tends to cold flow, it is sometimes necessary to retighten these joints.

8.2 STOPCOCKS

A. Vacuum Stopcocks. The proper selection and maintainance of stopcocks is important because they are a primary source of leaks in vacuum or inert-atmosphere systems. Close tolerances are necessary, and therefore most high-vacuum stopcocks have the plug and shell individually lapped together and numbered to avoid mismatch. Hollow-plug vacuum stopcocks seem to give the best performance. The common patterns for such stopcocks are shown in Fig. 8.7. For those shown in Fig. 8.7a through 8.7d, the plug is held in the stopcock by the vacuum on the back side of the stopcock, thus reducing leakage. These types of stopcocks are only useful on vacuum systems. It is important to avoid

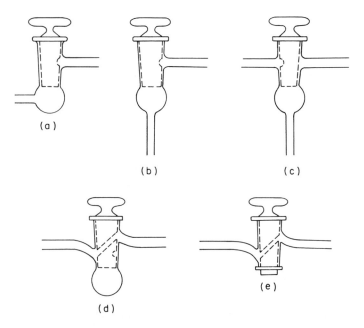

Fig. 8.7. Hollow-plug vacuum stopcocks. The hollow plug is firmly seated in types (a), (b), and (c) as long as the lower section is at reduced pressure. Stopcock (d) has a hollow plug with an oblique tube in the center; when the plug is turned by 180°, the lower section may be evacuated. Subsequently, only occasional evacuation of the vacuum cup is necessary. Type (e) does not have provision for evacuation of the plug and is therefore more liable to leak. Prolonged exposure of any of these stopcocks to solvent vapors erodes the grease and introduces leaks. (Adapted from Catalog C-64, Eck and Krebs Co., Long Island City, NY 11101.)

pressurizing these stopcocks, because the plug is then loosened and in an extreme case can become a projectile. It should be noted that the vacuum cup on the stopcock illustrated in Fig. 8.7d can only be evacuated through the lower sidearm. This factor requires that the stopcock be oriented with that arm pointing toward the source of vacuum.

Proper application of grease is even more important with stopcocks than with joints, because stopcocks generally have shorter leakage paths than do joints and the frequent actuation of the stopcock results in a more demanding situation. The barrel and plug should be clean and lint free. Four even, longitudinal stripes of grease are applied to the plug and it is inserted in the open position (Fig. 8.2). By application of pressure and small oscillations to the plug, the grease should form a continuous thin film. Large stopcocks may require evacuation to achieve a continuous film of grease. Permanent striations running around the stopcock and barrel may be due to minute particles of dirt or uneven or insufficient grease. For a reliable, vacuum-tight seal these striations must be eliminated; however, one must avoid the temptation to solve this problem by the use of excess grease. **NOTE: If a stopcock is greased too heavily, the grease extrudes into the orifice and clogs the opening. Furthermore, the thick film of grease will break down under vacuum by the formation of channels.** A guide to the choice of the optimum stopcock grease is given in Table 8.1.

It is best to evacuate a freshly greased system at least several hours before it is to be used and to work in the stopcocks by occasionally turning them. When high vacuum is required, it is usually advantageous to evacuate the system overnight and to flame the glass, exclusive of stopcocks and joints, to desorb moisture.

It is important to minimize the exposure of stopcocks and valves to volatile organic solvents at room temperature because the vapors dissolve in the stopcock grease. The presence of solvent vapors accelerates the tendency for stopcock grease to be squeezed out of the stopcock; and on long standing in the presence of solvent vapors, the stopcock grease will literally wash out of the joint or stopcock. Once the grease has absorbed solvent vapors, high vacuum may be unattainable because the solvent vapors outgas for a long time.

When the grease film has become extremely thin, either by the influence of long standing or of solvents, the leak-free operation of the stopcock will be impaired. This is generally evident by the appearance of striations in the stopcock grease or by the sudden cloudy appearance of the grease film when the stopcock is turned. In this event, thorough cleaning and regreasing are in order. When the grease becomes too thin, it is possible for the stopcock to stick in the shell. A seized stopcock can be freed by heating the shell with an air-gas flame and pulling the plug out with a twisting motion. This operation should be carried out in a smooth, rapid sequence. A bushy medium- hot flame is used and the stopcock shell is heated evenly for perhaps 15 seconds. The shell is grasped with a cloth towel or gloved hand to stabilize it and the plug is simultaneously turned and pulled out with the other hand. As described for joints in Section 8.1.A, stopcocks should be thoroughly cleaned before they are regreased and reassembled.

B. Lapping Stopcocks. Even though commercial vacuum stopcocks may be lapped by the manufacturer, it is occasionally necessary to repeat the process in the laboratory. A stopcock for which a striation-free film of grease cannot be attained, or one in which the grease consistently fails prematurely, are candidates for relapping. In the extreme case, a poorly matching plug and shell will be evident from a tendency for the plug to bind when it is turned. Sometimes stopcocks distort when they go through an annealing oven, and this is very often the case when a repair is made very close to the stopcock. Finally, if one member of a lapped pair is broken, the substitute part will have to be lapped with the remaining part. Before the operation is started, grease must be thoroughly removed from both parts, not only on the mating surfaces but also in the orifices. To provide an indication of the progress of the grinding operation, the plug and barrel are both marked with four or more pencil lines running the length of the ground glass.

If the stopcock is badly warped, it is necessary to start with coarse grit, such as 150-mesh corundum. A small portion of grit is made into a paste with glycerin or a 50–50 mixture of water and glycerin. This thin paste is applied to the plug in a smooth, even layer. The glycerin is not essential, but it is highly recommended because it reduces the tendency for the parts to seize. The two parts are ground together by an oscillating motion; the plug is then removed, turned by about one-fourth of a revolution, and replaced. The grinding is continued with care to replenish the grit and lubricant before the plug becomes dry. If the parts become dry or are forced, the plug may seize in the shell and ruin the job. Progress of the rough lapping should be followed by noting the disappearance of the pencil marks. Excessive grinding with the coarse grit must be avoided because this causes a mismatch of the orifices in the plug with those in the shell. When the rough grinding is complete, the parts are washed thoroughly to remove the last traces of grit, and the roughness of the surfaces is reduced by repeating the process using a fine grit and the water-glycerin lubricant. Either 600-mesh corundum or abrasive polishing alumina is satisfactory for the finishing operation. The latter has the advantage that it can be removed easily and thoroughly from the apparatus by a quick sodium hydroxide rinse followed by water and acid rinses.

C. Stopcocks for Pressure and Vacuum. In a typical Schlenk system, the apparatus is repeatedly exposed to a small positive pressure and to vacuum. This causes stopcocks to become dislodged; therefore, satisfactory performance requires good stopcock retainers. An excellent retainer design is illustrated in Fig. 8.8. This particular retainer requires the use of the stopcock produced by the same manufacturer.[6]

---

[6]An excellent stopcock and retainer design for use with Schlenk systems is available from Kontes Glass Co. (address in footnote 1).

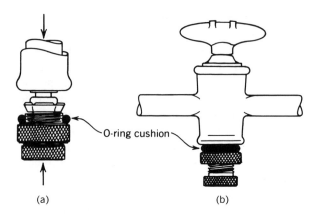

(a)   (b)

Fig. 8.8. Retainer stopcocks for vacuum and pressure. This style of stopcock (Kontes Inc.) is especially useful on a Schlenk manifold where both vacuum and small positive pressures of inert gas are used. (*a*) The retainer is being snapped onto the base of the stopcock plug. (*b*) The knurled nut has been tightened to hold the plug in place.

D. Metering with Stopcocks. Stopcocks are not very satisfactory for regulating the flow of gases, but improved flow control can be achieved by notching the orifices on the plug, as illustrated in Fig. 8.9. This is accomplished by drawing the corner of a three-cornered file across the orifice in the direction which is at right angles to the plug.

8.3 VALVES

A. Glass Valves. A few of the many different styles of glass valves on the market are illustrated in Fig. 8.10. The general design consists of a glass body housing either a glass or plastic (Teflon or Kel-F) stem. When the stem is con-

Fig. 8.9. Notched stopcock plug. A triangular file drawn across the orifice at right angles to the length of the plug will notch the orifice. This may then be used to better control gas flow, such as an inert-gas flow while working on a Schlenk manifold.

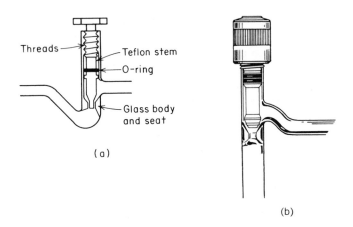

Threads

Teflon stem

O-ring

Glass body
and seat

(a)

(b)

Fig. 8.10. Teflon-glass valves. A number of variants of this basic design are on the market. A "straight through" flow pattern shown here for valve (a); (b) is a "right-angled" design. Type (a) uses a threaded Teflon stem working in a threaded glass body. In type (b) the stem does not rotate, so when the cap and Teflon nut are turned, the stem is forced up or down. Both styles are available with an O-ring on the tip of the stem. Also, some manufactures offer extended tips, shown in (a), which permit gas flow control. (Illustration (b) reproduced by permission of the copyright owner, Kontes Inc.)

structed from Teflon, a direct Teflon-to-glass seal provides good gas cutoff at the seat. The seal between the glass housing and the stem is generally achieved with O-rings, although there are at least two designs on the market (Quick-Fit and Young) in which the seal is Teflon-to-glass. Although they are generally not as leak-tight as conventional stopcocks, these grease-free valves have the tremendous advantage that they may be used in the presence of liquid solvents or with solvent vapors close to their condensation temperature. The Stock mercury float valve, which formerly provided the most satisfactory glass vacuum valve for use in the presence of solvents, has been almost completely replaced by these newer glass valves.

In several valve designs, the seal at the seat is made by means of an elastomeric O-ring on the tip of the valve stem. Leak tests indicate that this design offers no improvement in sealing over the valves in which there is a direct Teflon-to-glass seal at the seat. Furthermore, an O-ring at this point is vulnerable to solvents; so it is the opinion of the authors that a direct Teflon-to-glass seal at the seat is preferable for general use. When a valve is subject to temperature fluctuations, an O-ring on the tip of the stem is an advantage. Teflon has a much higher expansion coefficient than glass, so a valve with a Teflon stem ordinarily should not be cooled. If cooling is necessary, the seal at the seat can be maintained by tightening the valve as it cools. If the valve has an O-ring on the seat, it should not be cooled below the brittleness temperature of the elastomer.

The major leakage path in these valves is past the seal between the glass body and the valve stem, and the various products on the market vary widely in their leak rate at this point. A visual indication of a potential leaky valve is poor O-ring contact with the shell or, in those designs using a Teflon contact, uneven contact between the Teflon and the glass body. In general, it is desirable for the O-ring to be under high compression against the glass body. If there is only a thin line of contact, the chances are great that the valve will leak badly. Some improvement in performance is achieved with a light coat of stopcock grease on the O-ring, but this measure is defeated in the presence of solvents. Most valve designs provide protection of the O-ring seal on the stem by means of a Teflon wiper machined into the stem, or a Teflon O-ring adjacent to the elastomeric O-ring.

Since the Teflon-to-glass seal at the seat of the valve is much more dependable than the O-ring-to-glass seal at the stem, the valve should be attached to a piece of apparatus in the orientation which best utilizes the seat seal. For example, if the valve shown in Fig. 8.10a is used on a portable gas bulb, the seat seal on the valve should be closest to the bulb. Any leakage into the bulb would then have to occur past the seat seal. If the bulb were attached to the other arm on the valve, the bulb would be exposed to the relatively poor stem seal. Any leakage past the stem seal would then contaminate the contents of the gas bulb.

B. Metal Valves in Glass Apparatus. The excellent performance of metal diaphragms and metal bellows valves has prompted their use in glass systems. The various types of metal valves are discussed in detail in Chapter 10; therefore, this section will address the problems of compatibility of metal valves and glass apparatus. The union of metal valves to glass apparatus can be achieved easily by means of the swage-type connectors described in sections 8.1.C and 10.3.C. Because metal valves are generally heavy and more robust than the glass tubing they are connected to, the metal valves may imbalance a

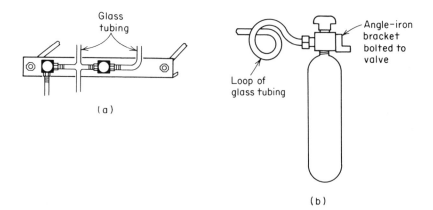

Fig. 8.11. Mounts for metal valves used in conjunction with glass systems. (*a*) The valves are rigidly mounted on a metal framework. (*b*) A semiflexible loop of glass is included to reduce mechanical strain with bulky parts.

glass apparatus and there may be a tendency to put too much stress on the glass tubing when the valve is installed or actuated. The best general approach to minimize these problems is to bolt the valve to a metal bracket, which in turn is mounted solidly to the vacuum rack. Another measure to reduce the transmission of stress to the glass parts is to use a semiflexible connection to the glass system by means of a short length of metal bellows tubing or a loop of small-diameter glass tubing (Fig. 8.11).

9

SPECIALIZED VACUUM LINE EQUIPMENT AND OPERATIONS

In Chapter 5 the basic design and operations of a vacuum line were presented to provide a concise introduction to chemical vacuum line techniques. This chapter describes a variety of more specialized operations which may be consulted as the reader's needs dictate. Familiarity with the basic vacuum line operations described in Chapter 5 is assumed. Some of the topics presented here are used for the quantitative characterization of chemical reactions: tensimeters (Section 9.1.A) and tensimetric titrations (9.1.C); or properties of compounds: vapor pressures above room temperature (9.1.E), molecular weight determinations (9.1.F), melting points (9.1.H), and spectroscopic determinations (9.1.I.) Another group of techniques is employed for separations, either for isolation of products or for analytical purposes: low-temperature fractional distillation (9.2.B), vacuum line filtration (9.2.A), and gas chromatography (9.2.C). The long-term storage of gases and solvents is described in Section 9.3. Sealed tube reactions are described in Section 9.4. Finally, a pair of miscellaneous topics are given in Section 9.5: sampling alkali metals (9.5.A) and generation of dry gases (9.5.B).

9.1 CHARACTERIZATION

A. Tensimeters. A tensimeter is an apparatus used to isolate a volatile sample on a high-vacuum system, and to characterize the sample by measuring the equilibrium vapor pressure. The tensimeter is used in a variety of important determinations, such as vapor pressure measurements on pure substances, gas-solution equilibrium measurements, gas-solution kinetic measurements, and related purposes such as molecular weight determinations and tensimetric titrations. Because of the wide variety of applications, scores of designs have appeared in the literature. This section provides three examples of tensimeter designs and discusses the main parameters which must be considered in the design of new systems.

The apparatus illustrated in Fig. 9.1 provides a simple, portable, and grease-free system for the wide variety of applications mentioned in the introduction to this section. To achieve portability, this apparatus is built with approximately 10-mm-tubing, which enables the determination of pressures with acceptable accuracy for most purposes and yet limits the volume of mercury so that the

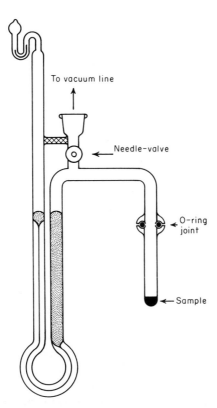

Fig. 9.1. Multipurpose tensimeter. Since the standard taper joint is positioned close to the manometer, the joint is close to the balance point of the apparatus. The tensimeter is best supported by clamping at the taper joint and by resting the bottom of the tensimeter on a piece of foam in a small dish which in turn is supported by a ring clamp. The brace between the left arm of the manometer and the joint is held in place with epoxy cement and adds stability to the apparatus. The tensimeter is filled with mercury in the same manner as a U-manometer (see Fig. 7.1 caption). To minimize the volume of mercury, capillary tubing is used for the bottom and part of the left leg.

apparatus is easily portable. The demountable sample tube permits the introduction of nonvolatile solids. This sample tube is attached to the apparatus with an O-ring joint (rather than a greased joint), thereby permitting long-term measurements in the presence of solvent vapors. The apparatus is equipped with a Teflon-glass needle valve with the seat toward the interior of the tensimeter. As explained in Section 8.3.A this orientation of the valve reduces leakage into the interior of the tensimeter. Also note that the main standard taper joint is positioned close to the manometer, and thus close to a point of balance for the apparatus. This feature, as well as the cross brace, make the apparatus reasonably sturdy and easy to handle.

The volume inside this tensimeter varies with pressure. Therefore, in situations which require a knowledge of the number of moles of gas, it is necessary to construct a calibration curve relating the internal volume of the tensimeter to the mercury level. The procedure for calibrating a manometer and attached volume has been described in Section 5.3.H.

Another type of tensimeter permits the mercury level to be manipulated so that a constant gas volume is achieved. One such constant-volume tensimeter is illustrated in Fig. 9.2. This apparatus is sufficiently unwieldy that it is best treated as a permanent item of equipment, with the mercury reservoir well supported by a fixed plaster base and the inlet tube fused to the vacuum system. In the version shown here, the sample tube is also fused to the apparatus, thus reducing leakage which is inevitable when joints are used. Owing to the lack of leakage paths, this apparatus is useful in situations where very long equilibration times are necessary. Since the apparatus is fixed, it is possible to use large-diameter tubing in the manometer and thus obtain great accuracy in the pressure measurements, if this is necessary.

Vapors can be transferred into or out of the apparatus by applying a rough vacuum to the reservoir which draws mercury out of the U and into the reservoir. After the volatile materials are condensed into the apparatus, they are isolated from the vacuum system by bringing the mercury reservoir to atmospheric pressure and slowly bleeding mercury into the U. If it is important to know the gas volume in the apparatus, the mercury level is adjusted to some reference point, such as A in Fig 9.2, for which the volume of the apparatus has previously been determined.

Electronic manometers provide a convenient method of pressure measurement in a tensimeter, and the general arrangement may be very simple (Fig. 9.3). The one problem which must be anticipated is long-term zero pressure drift, which can be encountered with an electronic pressure gauge. Drift is minimized by maintaining a constant temperature on the pressure transducer and by avoiding mechanical vibration at the transducer.

B. Thermal Transpiration.   In most tensimetry, it is correctly assumed that the pressure at the manometer is equal to the pressure in the sample container to

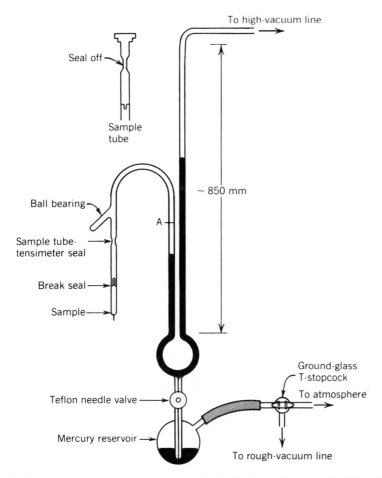

To high-vacuum line

Seal off

Sample tube

Ball bearing

Sample tube-
tensimeter seal

Break seal

Sample

~ 850 mm

A

Teflon needle valve

Mercury reservoir

Ground-glass
T-stopcock

To atmosphere

To rough-vacuum line

Fig. 9.2. Constant-volume tensimeter. The sample tube (enlarged view, upper left) is loaded with solid in a dry box or by sublimation from the vacuum line. The tube is evacuated, sealed, weighed, then glassblown to the tensimeter. After evacuating and flame-drying the tensimeter, the mercury level is raised and the break-seal cracked. Since the mercury serves as the cutoff to the vacuum manifold, the sample is not exposed to grease, stopcocks, or joints. This design is desirable when very long equilibration times are necessary. The mercury level is adjusted at the volume-calibrated reference point, such as A on the diagram, if it is important to know the gas volume in the apparatus.

which it is connected. However, at low pressures, where the diameter of the connecting tubing is small relative to the mean free path of the gas molecules, the pressures in the two chambers are related by $P_a/P_b = (T_a/T_b)^{1/2}$.[1]

[1]Thermal transpiration is discussed in many books on vacuum technology. See, for example: S. Dushman, *Scientific Foundations of Vacuum Technique*, 2nd ed., rev. by J. M. Lafferty et al. (New York: Wiley, 1962).

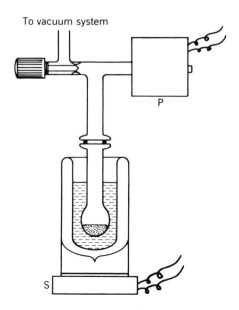

Fig. 9.3. Tensimeter based on an electronic pressure transducer. The sample with a microstirring bar is immersed in a constant-temperature bath. The pressure transducer (P), is connected to the glass apparatus by a glass-to-metal seal or a Swagelok fitting. S is a magnetic stirrer.

C. Example: Stoichiometry of Interaction by Tensimetric Titration.

This technique provides a convenient method for following the stoichiometry of a reaction between a gas and a solution or pure liquid. Also, similar techniques can be applied to the study of gas-liquid equilibria and gas-solid equilibria. In a tensimetric titration, measured amounts of a gaseous reactant are added to a solution. After each addition a pressure measurement is made. A plot of these pressures versus the moles of added gas is then inspected for the break-point, which represents the stoichiometry of interaction (Fig. 9.4).

A typical arrangement of components in a tensimetric titration is presented in Fig. 9.5, which shows the previously discussed tensimeter and a calibrated bulb attached to a vacuum line.[2] The sample container on the tensimeter is fitted with a small reciprocating stirrer which consists of a thin glass rod connected to a glass-encased headless nail or glass-encased bundle of soft iron wire. This stirrer is driven by an external solenoid, the field of which is switched on and off by a current-interrupting device, the details of which are laid out in Fig 9.6. The size of the calibrated bulb is chosen so that it will contain the desired amount of gas for each addition at a pressure which is convenient and accurately measured (e.g., 100–500 torr). The calibration procedure and steps used dispensing gas from such a bulb are described in Section 5.3.G.

[2]Examples of tensimetric titrations may be found in J. J. Rupp and D. F. Shriver, *Inorg. Chem.*, 6; 755 (1967).

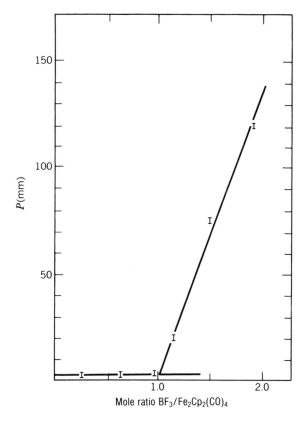

Fig. 9.4. Typical tensimetric titration curve. Total pressure in the tensimeter is plotted against the mole ratio of BF_3 added to reactant. In this case the solvent was toluene which was maintained at $-78°C$ for each pressure measurement. The horizontal portion of the pressure curve originates from the low toluene vapor pressure at this temperature. Above the 1:1 ratio of reactants, excess BF_3 is present and the pressure increases steadily with each addition of BF_3.

The sequence of operations (assuming the initial solid is not air sensitive) would be to load the sample tube with a weighed amount of reactive compound and the stirrer, to attach this tube to the tensimeter, and to pump out the air in the tensimeter. The sample tube is cooled to liquid nitrogen temperature and solvent is then condensed into the sample tube from a storage container on the vacuum line. The main valve on the tensimeter is then closed and the sample container allowed to warm so the solid may dissolve, perhaps with the aid of the stirrer. A constant temperature slush bath is next placed around the sample tube as illustrated in Fig. 9.5 and an initial pressure measurement is taken on the manometer. Next, the first alloquot of the reactive gas is transferred from a storage bulb elsewhere on the vacuum system into the calibrated bulb using the techniques outlined in Section 5.3.G (the bubbler manometer shown in Fig. 9.5 is used for the pressure determination required for this process). This gas is con-

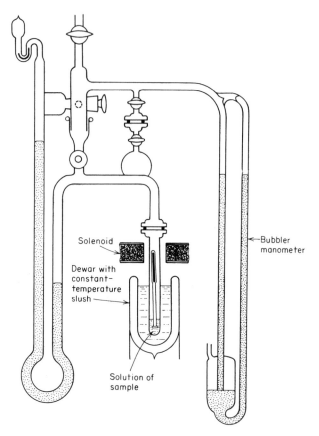

Fig. 9.5. Apparatus for a tensimetric titration. The bubbler manometer is used to measure aliquots of the reactive gas.

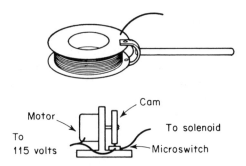

Fig. 9.6. Details of the solenoid stirrer. If the solenoid is wound with 2 lb of 22-gauge magnet wire, it will withstand 115 V in intermittent bursts. The current interrupter, shown above, involves a cam which opens and closes a microswitch.

densed into the sample tube, which has again been cooled to liquid nitrogen temperature, the contents of the tensimeter are isolated by means of the tensimeter valve, and the sample is again warmed to the slush bath temperature with stirring. When the pressure in the tensimeter has become steady, the pressure is recorded. Additions of gas and pressure measurements are continued in this fashion, and eventually a plot of pressure versus moles of added gas can be constructed as illustrated in Fig. 9.4. Often the equilibration times stretch out in the vicinity of the break-point if the reactants and/or products are only partially soluble. Slow equilibration times also result from slow gas diffusion when the vapor pressure of the solvent is great. In general, it is best to choose the solvent and the temperature for the measurement so that the solvent vapor pressure is low.

D. Example: Stoichiometry of Interaction by Differential Gas Absorption Measurement.

The greatest utility of the tensimetric titration is in observing reactions which are favored at low temperatures. If, however, the product has a negligible dissociation pressure at room temperature, it is simpler and often quite satisfactory to add an excess of reactive gas to the solution in the tensimeter, to stir the solution, and to monitor the pressure until no more gas is absorbed. The solvent and excess reactive gas can then be condensed into the vacuum system, separated by trap-to-trap distillation as described in Section 5.3.D, and excess reactive gas measured by its pressure, volume, and temperature in a trap on the vacuum line (see Section 5.3.H). Another check on the stoichiometry of the interaction is provided by the weight gain of the sample tube and its contents. Assuming that the product is air sensitive, the weight after reaction is determined by filling the tube with nitrogen, quickly removing it from the tensimeter, capping it with a small tared plug, and weighing the tube and plug.

E. Vapor Pressures above Room Temperature.

Since a volatile liquid will distill to the coldest point in an apparatus, it is necessary to thermostat the entire tensimeter system when vapor pressures are determined above room temperature. Two different designs are presented in Fig. 9.7 which meet this requirement; alternatively an immersible glass Bourdon pressure transducer may be used. The apparatus in Fig. 9.7.b is suitable for the measurement of gas-phase equilibria as well as vapor pressures. The first and simplest design of the two (Fig. 9.7.a), called an isoteniscope,[3] is operated in the following manner: On a vacuum system, liquid is condensed into the terminal bulb. A few hundred torr of an inert gas is introduced, the valve is turned off, the apparatus removed from the vacuum system, and the frozen liquid is allowed to melt. The part of the liquid in the terminal bulb is now tipped into the lower U, and inert gas in the region between the bulb and the U is removed by gentle pumping on the system,

[3]A. Smith and A. W. C. Menzies, *J. Am. Chem. Soc.*, 32; 897, 907, 1412 (1910).

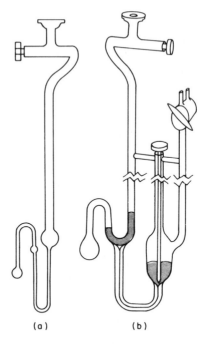

Fig. 9.7. Tensimeters for vapor pressures above room temperature. (*a*) Isoteniscope. (*b*) An immersible tensimeter.

with perhaps some warming of the bulb with one's fingers to promote gentle boiling rather than bumping. The apparatus is next attached to a manometer and surge bulb equipped with a stopcock through which inert gas or vacuum can be applied. The isoteniscope is immersed in a constant-temperature bath, and at each given temperature the liquid level on the two sides of the lower U is evened up by adjusting the inert-gas pressure in the external manometer-surge bulb combination. When a null point is reached for the liquid levels in the lower U, the vapor pressure of the liquid is indicated by the pressure on the external manometer. The presence of the inert gas in this manometer system prevents the rapid distillation of liquid out of the isoteniscope.

The apparatus in Fig. 9.7.*b* is the simplest to operate because the entire manometer system, the sample, and its vapor are immersed in the constant-temperature bath. The mercury reservoir permits the removal of mercury from the U-manometer portion of the apparatus, and the material to be measured is then condensed into the terminal bulb. While this material is still condensed and there is a high vacuum in the system, the mercury is reintroduced to the U, thus isolating the sample. The apparatus is immersed in a constant-temperature bath to the level of the wavy lines. With a vacuum on the upper portion of the apparatus the vapor pressures can be measured directly. If vapor pressures beyond the range of the immersible manometer must be measured, the mercury in the thermostated apparatus can be used as a null indicator as described above, with an external manometer and supply of gas connected at the upper joint.

Many other methods for vapor pressure measurements may be found in the literature and in general these may be adapted for air-sensitive compounds.[4-7]

F. Molecular Weight Determinations.

In principle, any standard method of molecular weight determination may be adapted for use under air-free conditions, and those which involve inert-atmosphere techniques have been covered in Section 1.5.F. This section will cover molecular weight determinations which require the use of a vacuum system.

Vapor density provides a useful means of molecular weight determination for substances which exert a high vapor pressure at or near room temperature. In the simplest case, an evacuated bulb of known volume is tared, filled with a measured pressure of the gas at a known temperature, and is reweighed. These pressure, temperature, and volume data yield the moles of gas (if the ideal gas law is assumed) and the additional weight data permit the calculation of the number of grams per mole.

The high surface area of the bulb and small weight of the gas reduce the accuracy of this determination because the large surface can pick up significant weight by adsorption of moisture from the atmosphere and from fingerprints. To minimize these sources of errors, the bulb may be wiped with a lint-free cloth moistened with water or alcohol before each weighing. The inaccuracy introduced by the adsorption of atmospheric moisture on the surface of the bulb can be minimized by using another bulb of the same size as a counterweight and treating both bulbs in an identical manner. The two bulbs are allowed to equilibrate with the atmosphere in the double-pan balance enclosure for about one-half hour before weighing.

Liquid samples can be weighed in a small, grease-free bulb and then transferred completely to an apparatus such as the one shown in Fig. 9.7.*b*, where the sample is completely volatilized and the pressure, volume, and temperature measured at an elevated temperature. Other methods for volatile substances include the gas density balance[8] and effusion measurements[9].

The molecular weights of nonvolatile substances which have a high solubility in a solvent with a high vapor pressure are conveniently measured on the vacuum

[4]D. E. McLaughlin and M. Tameres, *J. Am. Chem. Soc.*, 82; 5618 (1960); D. E. McLaughlin, M. Tameres, and S. Searles, Jr., *J. Am. Chem. Soc.*, 82; 5621 (1960).

[5]G. W. Thompson and D. R. Douslin, in A. Weissberger and B. W. Rossiter, Eds., *Techniques of Chemistry*, Wiley-Interscience, 1971, Vol. 1, No. 5, p. 23.

[6]W. Swietowlawski, *Ebulliometric Measurements*, (New York: Reinhold, 1945).

[7]G. W. Thompson and D. R. Douslin, in *Techniques of Chemistry*, A. Weissberger and B. W. Rossiter Eds., Wiley Interscience, 1972, Vol. 1, pt. 4, p. 23.

[8]H. L. Simons, C. L. Scheirer, and H. L. Ritter, *Rev. Sci. Instr.*, 24; 36 (1953).

[9]L. K. Nash, *Anal. Chem.*, 20; 258 (1948); D. P. Schoemaker and C. W. Garland, *Experiments in Physical Chemistry*, (New York: McGraw-Hill, 1962), p. 102.

line by the vapor pressure depression of the solvent. Assuming the solution be-
haves ideally, Raoult's law can be used to calculate the mole fraction of solute in
the solution:

$$P = P°X_{solvent} \qquad X_{solute} = 1 - X_{solvent}$$

In principle, only the weights of the solute and solvent used to prepare the solution, and the vapor pressure of the solution P, would have to be determined at a
known temperature. To reduce errors, the vapor pressure of the pure solvent $P°$
is simultaneously determined, with both solution and pure solvent tubes thermostated in the same constant-temperature bath. Very good data can be obtained
by using two of the tensimeters illustrated in Fig. 9.1. A more specialized apparatus, which utilizes a differential manometer, is shown in Fig. 9.8. To obtain
the best results using either apparatus, it is important to make careful measurements with a cathetometer, and to employ as high a temperature as practical so

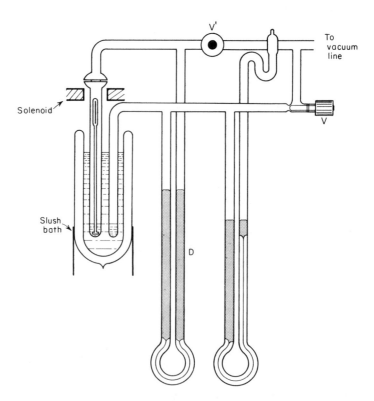

Fig. 9.8. Apparatus for the determination of molecular weights by vapor-pressure depression. The
bottom of the differential manometer (D) may be frozen with Dry Ice to keep the mercury from
sloshing when solvent is added or removed, or when the sample tube is attached. The solenoid used
to stir the sample tube is described in Fig. 9.6. V and V′ are grease-free valves.

that $P°$ is large and the difference between P and $P°$ is magnified. Under these conditions, it is quite important to recognize that a correction may be necessary for the quantity of the solvent in the gas phase, since for high temperatures and large volumes the amount of solvent in the solution may be significantly reduced.

For molecular substances, a wide variety of solvents are useful—for example, methylene chloride, pentane, dimethyl ether, and trimethylamine. Liquid ammonia is also frequently used. With 1:1 electrolytes, ammonia has the advantage that the primary species are ion pairs, whereas solvents of lower dielectric constant generally lead to higher and less definite degrees of aggregation. Correction for nonideality can be made by determining the apparent molecular weight as a function of concentration and extrapolating to infinite dilution.

The isopiestic (isothermal distillation) method for the determination of molecular weights is closely related to the vapor pressure depression method.[10] A weighed amount of standard is introduced into one leg of an apparatus and a weighed portion of the unknown is placed in the other leg. Solvent is introduced into the apparatus, which is then evacuated and thermostated. The solvent will distill from one solution to the other until the vapor pressures (and therefore mole fractions) of the two have equalized. If the solutions are ideal, or if the deviations from ideality are similar, equilibrium will occur when the mole fraction of the known equals that of the unknown.

A simple isothermal distillation apparatus is illustrated in Fig. 9.9. While not specifically designed for air-sensitive compounds, the apparatus may be loaded in a dry box. After introduction of samples and solvent, it is sealed off under vacuum and allowed to equilibrate, with the apparatus arranged so the liquid is in the large bulbs in order to afford a maximum area of exposed solvent. The apparatus is periodically tipped so the two solutions flow into the calibrated legs. When no change in volume is observed from one time to the next, the relative volumes of the solutions are, to a good approximation, proportional to the moles of solute in the two legs.

Since equilibration sometimes takes days or weeks, the isopiestic method is not suitable for unstable compounds or compounds with appreciable volatilities. Also, long equilibration times require leak-free and grease-free apparatus. Another design for an isothermal distillation apparatus is given in Fig. 9.10.

G. Temperature Determinations.   In vacuum line work it is frequently necessary to measure low temperatures. While the mercury-in-glass thermometer is convenient, it does not extend below the freezing point of mercury, $-38.9°$. Pentane-in-glass thermometers have a similar convenience and may be used down to about $-150°C$, but they are only useful for rough work (an error of $5°C$ is common). For more precise determinations, a calibrated thermocouple or thermistor, or a vapor pressure thermometer, is useful.

The vapor pressure thermometer consists of a manometer with one arm ex-

[10]G. L. Beyer, in *Techniques of Chemistry*, A. Weissberger and B. W. Rossiter Eds., Wiley Interscience, 1972, Vol. 1, pt. 4, p. 183.

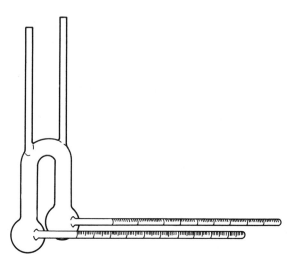

Fig. 9.9. Signer-type isopiestic molecular weight apparatus. The sample, standard, and solvent are introduced through the upper tubes, which are then sealed off after freezing the solutions and evacuating the apparatus. The apparatus is allowed to stand in the position shown to allow maximum exposure of the solution to the vapor. Volumes of the standard and unknown solutions are found by tipping the apparatus so the calibrated legs are filled.

posed to the vapor of a pure substance which is condensed in the probe (Fig. 9.11). Providing that some condensate is present, the temperature of the probe is found by comparison of the observed vapor pressure with tabulated data for the thermometric material. The vapor pressure thermometer illustrated in Fig. 9.11 is similar to Stock's design.[11,12] More compact designs have appeared in the literature,[13,14] but some sacrifice in accuracy seems inevitable with these.

Any substance which may be obtained pure and for which accurate vapor pressure data are available may be used for vapor pressure thermometry. Stock recommended a convenient series of compounds and, in collaboration with Hennig and others, he obtained accurate vapor pressure data for these compounds.[15] These data were obtained on a temperature scale for which $0°C = 273.1K$; however, more recent data on several of the compounds show it would be difficult to improve on Stock's vapor pressure tables. When high accuracy is desired, the usual precautions involved in precise manometry (Chapter 7) must be observed.

Thermistors, or "thermally sensitive resistors," are semiconductors which have high negative temperature coefficients of resistance. There is no simple re-

[11]A. Stock, *Z. Elektrochemie*, 29; 354 (1923).

[12]A. Stock, *Hydrides of Boron and Silicon*, (Ithaca, N.Y.: Cornell University Press, 1933), p. 190.

[13]A. Farkas and L. Farkas, *Ind. Eng. Chem. (Anal. Ed.)*, 12; 296 (1940).

[14]S. C. Liang, *Rev. Sci. Instr.*, 23; 378 (1952).

[15]F. Hennig and A. Stock, *Z. Physik.*, 4; 227 (1921).

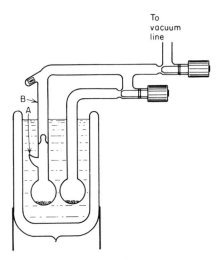

Fig. 9.10. Apparatus for the determination of molecular weights of air-sensitive compounds by the isothermal distillation technique. Temperature fluctuations of the two solvents are minimized in this illustration by a Dewar filled with water. This apparatus is used in the following manner. In a dry box the sample is placed in a tared tube through sidearm A, the sidearm is then sealed off, the tube and remnant of the sidearm are weighed, and the tube is attached to the apparatus by glassblowing at B. A weighed portion of a standard is introduced into the other bulb, and the filling tube is sealed off. After evacuation, opening of the break-seal, and reevacuation, a measured portion of solvent is distilled into both arms of the apparatus. The process of equilibration is followed by periodic removal and measurement of the solvent from one arm. The solvent may be measured volumetricly in the liquid or gas states, or by weight.

lationship between the resistance R and the temperature T. However, the following empirical resistance/temperature relationship is quite useful:[16]

$$R_T = R_o \exp\left[-B\left(\frac{1}{T} - \frac{1}{T_o}\right)\right]$$

where R_o is the resistance at a temperature T_o, R_T is the resistance at a temperature T, and B is a constant.

Low-temperature thermistors are usually made from nonstoichiometric iron oxides and have a resistance sensitivity of around 15% per Kelvin at 20 K.[17] Thermistors act as ohmic conductors at any fixed temperature. Therefore, one advantage of using a thermistor is that ordinary copper wiring may be used to build the circuit; reference junctions and special extension wires are not needed.[18] Thermistors are generally quite stable to long-term fluctuations after an initial aging period.

[16]S. D. Wood, B. W. Mangum, J. J. Filliben, and S. B. Tillett, *J. Res. NBS*, 83; 247 (1978).

[17]T. J. Quinn (general references), p. 223.

[18]R. P. Benedict (general references).

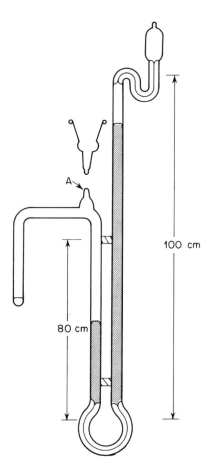

Fig. 9.11. The vapor-pressure thermometer. The manometer is filled as described in the caption to Fig. 7.1. The manometer is then evacuated and outgassed through a joint attached at point A. A pure compound is condensed into the probe, and the apparatus is sealed off under vacuum at point A. For thermometric compounds which boil below room temperature, care must be taken not to overfill the apparatus.

Thermocouples consist of two dissimilar electrical conductors which are joined to form a measuring junction, with the free ends of the wires constituting the reference junction. When a temperature difference exists between the measuring and reference junctions, an emf is produced between the free ends of the device. This emf, which is a function of the temperature difference, can be used to determine the temperature at the measuring junction if the reference junction temperature is known. A schematic of a typical thermocouple circuit is shown in Fig. 9.12.

The most common thermocouple type used in a laboratory situation is the Type J iron/constantan thermocouple. Low cost, high thermopower, and a use-

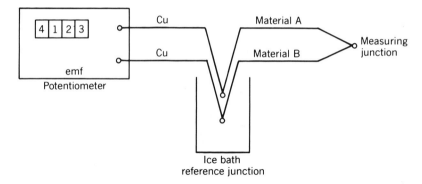

Fig. 9.12. Typical thermocouple circuit. The ice water bath may be prepared in a Dewar for extended use of the circuit.

ful temperature range of $-200-760°C$ contribute to the popularity of the Type J thermocouple. Below $0°C$, rust and embrittlement may become a problem.[19] Therefore, it is a good practice to remove the reference junction from the ice water bath and dry it thoroughly when the thermocouple is not being used.

When using a measuring junction at a remote site relative to the reference junction, it is important to use the proper extension wire between the junctions. Usually, a wire with the same composition as the thermocouple wire itself, but not made to such a high specification, is used.

H. Melting Point Determinations. The melting point, or more correctly the triple point, of a volatile solid may be determined in a standard melting point block or bath after the solid is sublimed from the vacuum line into a capillary and sealed off under vacuum. To avoid sublimation of the sample during the determination, the capillary must not protrude from the melting point block or bath.

Stock devised the apparatus illustrated in Fig. 9.13 to obtain melting points of volatile and reactive substances below room temperature. This particular apparatus overcomes the problem of poor visibility of the sample, which is being maintained at a low temperature. The apparatus contains a glass plunger which has an iron core on the top and a glass cross at the bottom. The plunger is slipped into a tube which allows moderate clearance. In operation, the tube with plunger is evacuated, the plunger is lifted by a hand magnet, and a ring of the sample is distilled into the apparatus below the cross. The plunger is lowered so that the cross rests on the frozen ring, and a bath of cold liquid is placed around the tube and allowed to warm slowly. The temperature of the bath may be measured with a vapor pressure thermometer and the top of the plunger observed

---

[19]T. J. Quinn (general references), p. 264.

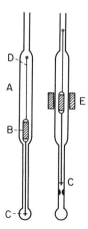

Fig. 9.13. Stock-type melting-point apparatus. A, tube; B, magnetic core; C, glass cross resting on the bottom (left drawing), poised above the ring of frozen solid (right drawing); D, tip of plunger which is observed; E, hand magnet used to raise plunger.

through a cathetometer. The melting point corresponds to the temperature at which the plunger begins to move.

Visual observation of melting points may be made in the apparatus illustrated in Fig. 9.14. The sample(s) and a mercury-in-glass thermometer are held by supports constructed from thin-walled brass tubes. The bath is cooled below the freezing point of the samples and stirred by a motor-driven stirrer. The apparatus may be placed in a warm water bath, and the rate of warming may be controlled by the extent of evacuation of the jacket. If the substance freezes as a solid phase which is difficult to distinguish from the liquid, it is illuminated from behind by light which is passed through a Polaroid film and then viewed through a crossed Polaroid.

I. Spectroscopic Determinations.   Gas-phase infrared spectra provide a useful adjunct to vapor pressure measurements in the identification of volatile materials. The cell illustrated in Fig. 9.15 allows the sample to be quantitatively returned to the vacuum line after the spectrum has been obtained, so the process is completely nondestructive. The primary problem with a gas cell is to obtain a vacuum-tight seal between the window material and the cell body; this may be accomplished with Glyptal paint or with wax. If the latter is used, it is necessary to warm and cool the alkali halide windows slowly to avoid cracking them due to thermal stress. For this purpose an infrared lamp is handy. The most satisfactory method of attaching windows is O-rings because this allows the easy removal of the windows for cleaning and polishing.

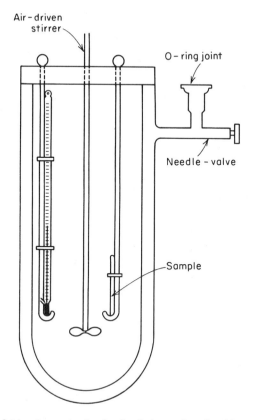

Fig. 9.14.   Apparatus for the visual observation of melting points.

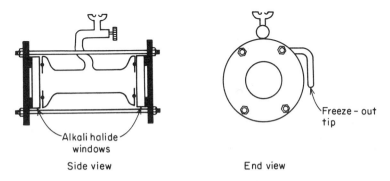

Fig. 9.15.   A gas-phase infrared cell. Note that a thin gasket is included between the metal retaining plate and the window. This helps to reduce window breakage owing to uneven tightening of the nuts.

A versatile low-temperature infrared cell is presented in Fig. 9.16.[20] It is suit-able for mull or pressed-disk infrared spectra of solids, solid state infrared spec-tra of condensed vapors, and spectra of products from solid-gas reactions. Mulled or powdered samples are supported between alkali halide windows in the copper block B, which is cooled by refrigerant in the cold finger A. Before this refrigerant is added, the cell must be evacuated through the needle valve. When the cell is used for low-temperature spectra of condensables, the vapors are bled through the needle valve and squirted directly onto one surface of a cold alkali halide plate held in block B.

The cell illustrated in Fig. 9.17 has proven very useful in the determination of visible and ultraviolet spectra for air-sensitive solutions. Despite its simplicity, this cell is quite versatile. To introduce an air sensitive solution, the cell is evacu-ated and filled with dry nitrogen through the O-joint, the stem of the needle valve is removed while a stream of nitrogen is maintained through the cell, the solution is injected by means of a syringe equipped with a long needle, and the stem is replaced. When the needle valve is closed, the solution may be tipped

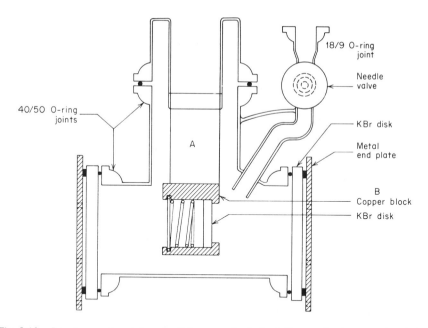

Fig. 9.16. Low-temperature infrared cell (cross-section). An indium gasket between the window and copper block greatly increases the efficiency of heat transfer. If the temperature of the window is to be measured with a thermocouple, the leads may be passed though pinholes in the upper O-ring. Under compression of the joint, this type of electrical lead-through is vacuum-tight. Threaded rods used to clamp the end plates have been omitted for clarity. A spring holds the KBr disk in place.

[20]This cell is similar to one described by E. L. Wagner and D. F. Hornig, *J. Chem. Phys.*, 18; 296 (1950).

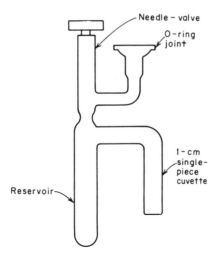

Fig. 9.17.   Evacuable cell for visible spectroscopy.

into the cuvette and the spectrum may then be run. If a measured portion of reactive gas is to be added to the solution, the contents are frozen in the reservoir, the cell is evacuated, and the gas is distilled into it. When this cell is used in conjunction with most visible-ultraviolet spectrometers, it is necessary to replace the usual cell compartment cover with a specially constructed, light-tight cover which provides space to accommodate the needle valve. The cylindrical reservoir is an important feature of the cell because the cuvette, which has a square cross section, is easily broken by frozen solutions.

Commercial cylindrical quartz cells can be adapted for gas-phase work as illustrated in Fig. 9.18. Such a cell finds use in the near infrared for the determination of overtone vibrational frequencies, and also in visible and ultraviolet spectroscopy. A much less expensive cell which is adequate for most gases may be constructed from Pyrex along the lines of the cell shown in Fig. 9.18. Quartz windows may then be attached by epoxy resin. A cell which is filled from a conventional vacuum line will generally contain mercury vapor which absorbs at 2537 Å. Once the origin of this absorption is recognized, it causes little difficulty because of its narrow bandwidth.

Many specialized spectral cells designed for reactive substances are described in the literature, and references to some of these are given in Table 9.1.

Laser Raman spectroscopy is well suited for the study of air-sensitive liquids because the sample may be contained in an all-glass cell.[21] Such a cell is much easier to load on a vacuum line and to maintain leak-free than is an infrared cell. Also, such a tube is easier to heat or cool than the typical IR cell.

[21]D. P. Strommen, and K. Nakamoto, *Laboratory Raman Spectroscopy*, Wiley Interscience, New York, 1984, Chapter 2.

Fig. 9.18. A quartz gas cell. The adapter with joint, stopcock, and freeze-out sidearm is waxed onto the taper joint of a commercial cell.

Several methods may be used to attach an NMR tube to a vacuum system. The O-ring union, illustrated in Fig. 8.5, permits attachment of standard NMR tubes to a vacuum outlet. Once the NMR tube has been filled with an air-sensitive sample on the vacuum system, it is necessary to bring the tube up to atmospheric pressure with inert gas, remove the tube, and quickly stopper it. Recently, a small in-line valve has been introduced which has a provision for an O-ring seal with a specially constructed NMR tube.[22] This valve allows the NMR tube to be sealed before it is disconnected from the vacuum line, and the valve is left on the tube during data collection. Owing to its symmetric design, the valve does not interfere with the spinner.

The best exclusion of air is achieved by glassblowing an NMR tube onto a joint so that the joint may be attached to the vacuum line when volatile compounds are to be sampled. The sample, solvent, and reference are distilled into the tube, which is then sealed off under vacuum. A symmetric seal helps to avoid erratic spinning. A highly volatile reference, such as tetramethylsilane, may be stored in a small "solvent container" (Fig. 9.19) and measured in the gas phase. The large expansion coefficient of many solvents can crack the sample tube when the sample is warmed from the liquid nitrogen temperature. To reduce the chance of breakage, the tube is withdrawn slowly from liquid nitrogen and, as it is removed, the sample is thawed with the fingers; it is particularly important to use this process with the thin-walled variety of NMR tube. The expansion coefficient also has to be taken into account so that there is sufficient dead space to accommodate the sample when it has warmed to room temperature. If high pressures are anticipated, a check is made of the strength of the tube by immersing it in a warm water bath before it is inserted into the NMR probe.

---

[22]The in-line valve and associated NMR tube is manufactured by J. Young Ltd., 11 Cloville Road, Acton, London W38BS, U.K. In the United States one supplier is R. J. Brunfeldt Co., P. O. Box 2066, Bartlesville, OK 74005.

Table 9.1. References to Infrared and Visible-Ultraviolet Cells for Air-sensitive Compounds

Description	Reference
Infrared	
Vacuum-tight infrared cell for liquids	A. B. Burg and R. Kratzer, *Inorg. Chem.*, 1, 725 (1962).
Infrared cell for carbon suboxide polymer	R. N. Smith, D. A. Young, E. N. Smith, and C. C. Carter, *Inorg. Chem.*, 2, 829 (1963)
Pressure-tight cell for infrared spectra of liquids	D. C. Smith and E. C. Miller, *J. Opt. Soc. Am.*, 34, 130 (1944).
Method of attaching dissimilar window materials to a vacuum-tight low-temperature cell	E. Schwarz, *J. Sci. Instr.*, 32, 445 (1955); V. Roberts, *J. Sci. Instr.*, 31, 251 (1954)
Infrared spectra of adsorbed species	R. P. Eischens, S. A. Francis, and W. A. Pliskin, *J. Phys. Chem.*, 60, 194 (1956); M. Courtois and S. J. Teichner, *J. Catalysis*, 1, 121 (1965). C. Tessier-Youngs et al., Organometallics, 2, 898 (1983).
All-glass cell for the infrared ($<4.7\mu$) spectra of gases.	F. A. Cotton and L. T. Reynolds, *J. Am. Chem. Soc.*, 80, 269 (1958).
Low-temperature near-infrared and visible cells for liquid ammonia solutions.	W. J. Peer and J. J. Lagowski, *J. Phys. Chem.*, 84, 1110 (1980), E. C. Fohn, R. E. Cuthrell, and J. J. Lagowski, *Inorg. Chem.*, 4, 1002 (1965).
Heated infrared cell for solids in a controlled atmosphere	T. Wydeven and M. Leban, *Anal. Chem.*, 39, 1673 (1967).
Low-temperature liquid infrared cells	R. G. Steinhardt, P. A. Staats, and H. W. Morgan, *Rev. Sci. Instr.*, 38, 975 (1967)
Silver chloride window-to-body seals	J. F. Harrod and H. A. Poran, *Rev. Sci. Instr.*, 38, 1105 (1967); A. Guest and C. J. L. Lock, *Rev. Sci. Instr.*, 39, 780 (1968).
Optical materials and various infrared cell designs.	R. G. J. Miller, 1965, *Laboratory Methods in Infrared Spectroscopy*, Heyden and Sons Ltd., London.
Protective holder for KBr pressed disks	P. A. Saats and H. W. Morgan, *Appl. Spectry.*, 22, 576 (1968).
Visible-Ultraviolet	
Low-temperature vacuum-tight visible-ultraviolet cell for solutions.	R. Nakane, T. Watanabe, O. Kurihara, and T. Oyama, *Bull. Chem. Soc. Japan*, 36, 1376 (1963).
Low- and high-temperature gastight visible-ultraviolet cell.	Y. Hirshberg and E. Fischer, *Rev. Sci. Instr.*, 30, 197 (1959).

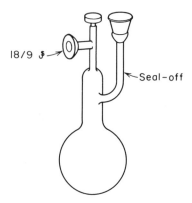

18/9 $\,\delta\,$ →

←Seal-off

Fig. 9.19. Solvent storage container. Solvent is distilled in through the sidearm, which is then sealed off under vacuum. The container is attached to the vacuum line through a 18/9 ball joint or O-ring joint.

Sampling for ESR work is often more complex than for NMR because the high sensitivity of the ESR spectrometer requires careful exclusion of paramagnetic impurities or their progenitors. Also, the radical is generally extremely reactive and present in low concentrations, so that exclusion of traces of moisture and air is important. Frequently, paramagnetic low-valent metal complexes and anion radicals are produced by alkali metal reduction; a typical cell for the generation and sampling of such species is illustrated in Fig. 9.20. A variety of other sampling methods have been described.[23]

9.2 SEPARATIONS

A. Vacuum Line Filtration.
For their work on the diammoniate of diborane, Parry, Schultz, and Girardot[24] devised a versatile vacuum line filtration apparatus which is useful when small quantities of solid are handled and when the solvent is sufficiently volatile to be distilled on the vacuum line. The filter is attached to the vacuum system through a standard taper joint which allows it to be rocked or inverted (Fig. 9.21). Prior to filtration, any volatile contents are frozen down and the apparatus is thoroughly evacuated (Fig. 9.21a). By inversion of the apparatus, the solution is then poured onto the frit, and the solvent vapor pressure is employed to effect a "suction" filtration by closing the stopcock in the equalizing arm and cooling the lower tube (Fig. 9.21b). The precipitate is washed by distillation of the solvent from the lower receiver into the upper portion of the apparatus (with the stopcock in the sidearm open) and repetition

[23]C. P. Poole, Jr., *Electron Spin Resonance*, (New York: Interscience, 1967), Chapter 15; J. L. Dye, *J. Phys. Chem. 84*, 1084 (1980).

[24]R. W. Parry, D. R. Schultz, and P. R. Girardot, *J. Am. Chem. Soc.*, 80; 1 (1958).

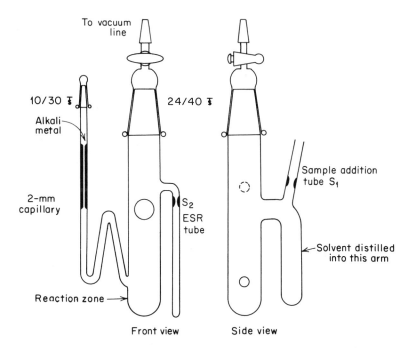

Fig. 9.20. Apparatus for the generation and sampling of anion radicals. The sample is added to the purged apparatus, S_1 is sealed off, and the appartus is evacuated. Chunks of sodium are melted through the capillary and then sublimed into the reaction tube. Solvent is distilled into the sidearm from the vacuum line, and the apparatus is tipped so that the resulting solution is poured onto the sodium mirror in the reaction zone. The resulting solution of radical is poured into the 5-mm ESR tube, frozen down, and sealed off at S_2. The radical generation and collection can be carried out at reduced temperature by immersing the tube in an appropriate slush bath.

of the suction filtration (Fig. 9.21*b*). The condensation of solvent in the upper portion of the filter can be accomplished by wrapping part of the upper tube with glass wool and pouring small quantities of liquid nitrogen into the glass wool. When the filtering and washing operations are complete, the solvent may be removed under vacuum. The apparatus is filled with nitrogen, capped, and taken into a dry box for the transfer of solids. This set of operations is easily carried out when using solvents which exert less than 1 atm pressure at room temperature; in this case the major precaution is slow removal of the solvent to avoid splattering of solid in the filtrate. However, solvents which boil below room temperature require much more skill in handling because they are liable to bump and blow the apparatus apart. The chance of bumping may be reduced by a rocking motion imparted to the apparatus when the solvent is volatilized. Also, pressurization of the apparatus is minimized by always leaving it open to a bubbler manometer. Even though the design presented in Fig. 9.21 avoids direct contact of the liquid with stopcock grease, the presence of solvent vapors may lead to the eventual erosion of the grease. This can be minimized by working

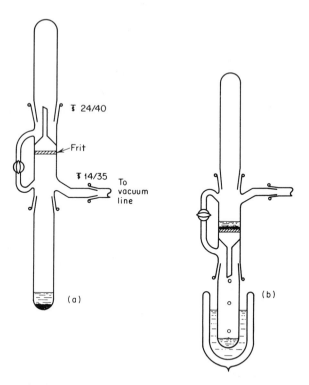

Fig. 9.21 Vacuum line filtration apparatus. (*a*) A solution ready to be filtered; (*b*) apparatus inverted and bottom receiver cooled to bring about suction filtration. The precipitate is washed by opening the stopcock, allowing the bottom tube to warm, and cooling the top with glass wool dipped in liquid nitrogen. The components are held together by springs or rubber bands. A grease-free version of similar dimensions may be constructed by making the following substitutions $\mathbb{T}$ 24/40 stopcock → valve; $\mathbb{T}$ 14/35 → 1/2-in. Cajon O-joint.

fairly rapidly and maintaining the solvent below room temperature, or by using the O-ring version of this apparatus, which is described in the caption to Fig. 9.21.

When a large quantity of solid is handled, a scaled-up version of the vacuum line filter may be used; but if the quantity is too large, it becomes unwieldy. The vacuum line filter design by Wayda and Dye, Fig. 5.3 is the best choice when large quantities are handled. Many specialized vacuum line filters are described in the literature.[25,26]

[25]G. D. Barbaras, C. Dillard, A. E. Finholt, T. Wartik, K. E. Wilzbach, and H. I. Schlesinger, *J. Am. Chem. Soc.*, 73; 4585 (1951).

[26]A. B. Burg and R. Kratzer, *Inorg. Chem.*, 1; 725 (1962).

B. Low-Temperature Fractional Distillation. A versatile low-temperature distillation column has been designed by Dobson and Schaeffer.[27] It is suitable for the separation of small amounts of compounds which have appreciable
volatilities in the −160–0°C range. The column (Fig. 9.22) is evacuated, and the
sample to be distilled is condensed in the lower U-trap. A stream of cold nitrogen
gas is maintained through the center tube, and the U-tube is warmed so that the
sample is introduced to the column. The temperature of the column is controlled
by varying the boil-off rate of liquid nitrogen from a Dewar (Fig. 9.23). When a
sufficiently high temperature is reached, the first component will slowly move up
the column and may be collected over a period of time at liquid nitrogen temperature in the upper U-trap. Ordinarily, a good separation is obtained for components which boil 15° apart. A number of other designs are presented in the literature for low-temperature distillation columns.[28,29]

A closely related technique is the "fractional codistillation" method of Cady
and Siegwarth.[30,31] The amount and boiling range of the components which may
be handled by this technique are comparable to those described above for the
low-temperature distillation column. Essentially, a gas chromatography apparatus is employed with the column replaced by a 1/4- or 3/8-in. copper U-tube
packed with approximately 50-mesh metal powder. The mixture to be separated, which may be as much as 2 mL of liquid for the large column, is distilled
onto the inlet side of the U-tube at liquid nitrogen temperature, and a suitable
helium flow rate is then established. The liquid nitrogen is poured from the Dewar and the Dewar is replaced around the U-tube. As the chilled Dewar and U-
tube warm up, the components progress through the column and then through a
thermal conductivity detector, which is followed by cold traps for the collection
of fractions. A simple version of this apparatus is illustrated and explained more
fully in Fig. 9.24.

C. Gas Chromatography. Gas chromatography has been employed for
the separation of reactive compounds ranging from boron hydrides to interhalogens. For many reactive liquids, conventional instrumentation (Fig. 9.25) has
been employed. However, gases generally require an inlet system which is more
involved than the familiar syringe injection port routinely used for organic compounds.

There is no lack of books and articles dealing with the major facets of gas
chromatography; for examples, see the general references list at the end of this
chapter. Furthermore, most chemists are familiar with the technique, and many

[27]A. Norman and R. Schaeffer, private communication.
[28]J. R. Spielman and A. B. Burg, *Inorg. Chem.*, 2; 1139 (1963).
[29]D. J. LeRoy, *Can. J. Chem.*, 28, 492 (1950); A. A. Comstock and G. K. Rollefson, *J. Chem. Phys.*, 19; 441 (1951).
[30]G. H. Cady and D. P. Siegwarth, *Anal. Chem.*, 31; 618 (1959).
[31]H. W. Myers and R. F. Putnam, *Anal. Chem.*, 34; 664 (1962).

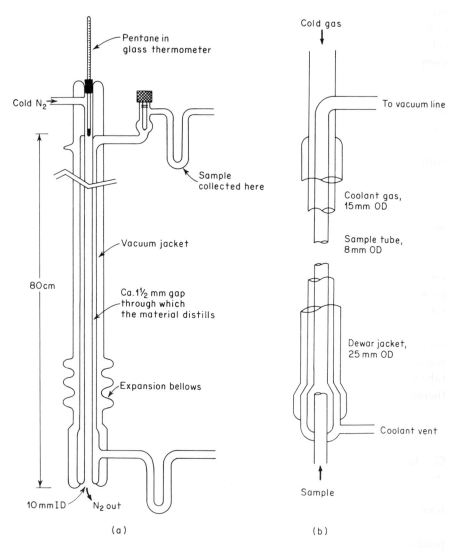

Fig. 9.22. Low-temperature fractional-distillation column. (*a*) Norman's modification of a design by Dobson and Schaeffer. (*b*) An alternate column design (J. Dobson, Ph.D. thesis, Indiana University, 1967).

commercial instruments are available. Therefore, this section will focus on inlet systems, particularly inlets which serve as an interface between a gas chromatograph and a vacuum system. Also, a brief account will be given of the conditions and equipment which have been used successfully in the separation of some reactive inorganic substances.

An inlet system for quantitative work must present a *representative* sample to the carrier gas stream in the chromatograph. This can be a problem with highly

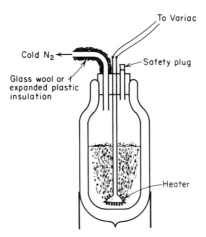

Fig. 9.23. Attainment of low temperatures by boiling nitrogen. In many applications such as the still in Fig. 9.22, the Variac is set to provide a convenient steady-state boil-off temperature. When close temperature control is desired, a commercial control with a thermocouple or thermistor sensor may be used to control the heat input.

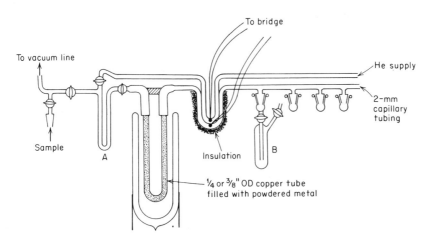

Fig. 9.24. Fractional codistillation apparatus. In this design (due to S. M. Williamson), the sample is distilled into U-tube A and then carried by a helium stream into the upper left leg of the metal tube, which is cooled with liquid nitrogen. When the liquid nitrogen is poured out and replaced by a chilled Dewar, the fractions are carried through the column by the helium stream and collected in trap B when they appear in the detector. (Only one trap for fraction collecting is illustrated.)

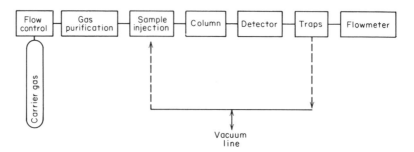

Fig. 9.25. Schematic of a gas chromatograph. The flow control generally consists of a standard pressure regulator plus a needle valve. Gas purification is sometimes necessary, particularly with air- and moisture-sensitive compounds. A short column of molecular sieves held at −78°C is frequently adequate. The traps (for the collection of fractions) and flow meter are optional. The dashed lines indicate points at which communication with the vacuum line is possible.

volatile substances, because partial condensation of a multicomponent system will lead to a different relative concentration of the components in the gas phase than in the liquid phase. The following procedure is recommended for obtaining a representative sample of a liquid containing highly volatile components:[32]

1. The liquid phase is made homogeneous,

2. A portion of the liquid phase is isolated from the bulk of the liquid without change in composition (i.e., without boiling).

3. The isolated portion is completely evaporated into a suitable container again without change in composition.

4. The vapor is made homogeneous.

Similarly, it is advantageous to obtain a gas sample from a homogeneous poly-component mixture by isolating a segment of the homogeneous gas phase by means of a valve(s) and then removing all of this sample for analysis.

Volatile reactive liquids present few problems since they generally can be distilled from the vacuum line into a tube equipped with a serum bottle cap (Fig. 9.26). Nitrogen is then admitted to the tube, and the sample is taken with a syringe. Another scheme involves an inlet with a capillary tube or ampule breaker.[33-35] This method is potentially useful for vacuum line work, since it is relatively simple to fill and seal off a sample tube attached to the vacuum line.

When slight contamination of a gas sample by air is not objectionable, syringe techniques may be used. In order to minimize contamination by air, a syringe

[32]V. Diebler and F. L. Mohler, *J. Res. NBS*, 39; 149 (1947).

[33]M. Dimbat, P. E. Porter, and F. H. Stross, *Anal. Chem.*, 28; 290 (1956).

[34]S. W. S. McCreadie and A. F. Williams, *J. Appl. Chem.*, 7; 47 (1957).

[35]J. Wendenburg and K. Jurischka, *J. Chromatog.*, 15; 538 (1964).

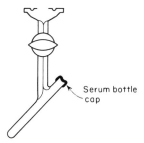

Fig. 9.26. Liquid sampling tube. The gas volume of this apparatus is kept small to minimize deple-tion of the more volatile component in the liquid phase.

with a Teflon- or rubber-tipped plunger is employed,[36] the syringe is flushed with carrier gas, and the pressure of the gas sample is adjusted to slightly more than 1 atm before the aliquot is withdrawn. While very convenient, this approach is not appropriate for highly reactive gases or situations in which rapid, accurate sampling is desired.

The common method of gas sampling is to employ a loop of tubing which may be cut in or out of the carrier gas stream. The loop is filled with a gas sample and then switched in series with the carrier gas source and column. The loop may be filled by evacuating it and then either admitting the sample at a measured pressure from the vacuum line or distilling the entire sample from the vacuum line into the loop (Fig. 9.27). Obviously, this process requires the use of valves which will hold a high vacuum. The valve requirements are less stringent when the sampling operation is performed by flushing the sample gas through the loop (Fig. 9.27). This approach is well adapted to the analysis of commercial gas streams and is useful in the laboratory for sampling the output of flow reactors. However, it is not convenient for most vacuum line work.

A wide variety of valves may be used to achieve the aforementioned gas injection. Special valves are available which provide synchronous switching of the sample into the gas stream. One type involves a body attached to the carrier gas line and sample loop(s). The passages in this body open on a flat, polished surface. A rotor with passageways which will interconnect those on the body is pressed firmly against it. By turning the rotor, the gas stream may be switched between different ports. One of the most satisfactory designs from the standpoint of leak-tight performance is based on a spring-loaded cone of graphite-impregnated Teflon which rotates in a metal body.[37] The porting is schematically illustrated in Fig. 9.28. Another type of valve involves a shaft bearing a

---

[36]S. H. Langer and P. Pantages, *Anal. Chem.*, 30; 1889 (1958). Teflon-tipped gastight syringes are manufactured by the Hamilton Company, Whittier, CA. "Disposable" syringes with rubber-tipped plungers are manufactured by Becton, Dickinson, Co., Rutherford, NJ.

[37]These valves are manufactured by Valco and are available from supply houses of chromatographic equipment, such as Supelco, Inc., Supelco Park, Bellefonte, PA 16823.

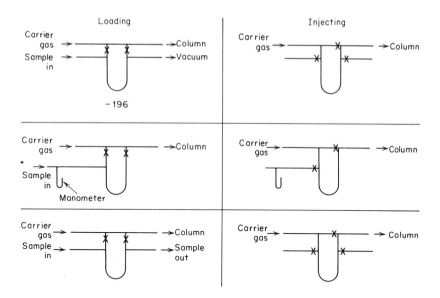

Fig. 9.27. Some configurations for gas sampling.

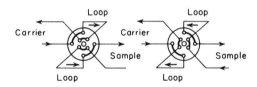

Fig. 9.28. Gas inlet valve. Schematic of cone-type or disk-type valve.

series of O-rings in grooves. This fits inside a cylinder from which passageways extend. Communication between the passageways is altered by sliding the O-ring bearing shaft in or out.

Before discussing some individual systems, it is worth pointing out that an ordinary analytical chromatograph will handle samples which are very close to the scale encountered in vacuum line work (up to 1 to 3 mmol with a 1/4-in.-diameter column). Therefore, the usual analytical chromatograph is potentially useful for the purification of products or reactants for small-scale vacuum line preparative work. For preparative work or for subsequent characterization of the fractions, the gas from the chromatograph is led through a small collection manifold where individual components are trapped.[38,39] Efficiency is improved by including loose glass wool or similar material in the trap.

[38]J. Haslam, A. R. Jeffs, and H. A. Willis, *Analyst*, 86, 45 (1961).

[39]C. M. Drew and J. R. McNesby, in Desty, Ed., *Gas Chromatography 1958*, (London: Butterworth, 1958), p. 216; G. Guiochon and C. Pommier, *Gas Chromatography in Inorganics and Organometallics*, Ann Arbor Sci. Pub., 1973.

Permanent gases may be separated by gas-solid chromatography. For example, oxygen, nitrogen, and hydrogen may be separated by a column packed with Linde 5A or 13X Molecular Sieves and detected by thermal conductivity.[40,41] Permanent gases such as H_2, O_2, N_2, CO, and CH_4, as well as carbon dioxide, terminal alkenes and straight-chain alkanes up to pentane are well separated by a temperature-programmed column consisting of a carbon molecular sieve called Spherocarb.[42] The separation of H_2, N_2, CO, CO_2, NO, and N_2O is possible on specially activated silica gel at reduced temperatures.[43] A more thorough review of the separation of light gases is given by Littlewood,[44] and by Thompson.[45] The fairly strong oxidizing agent NO_2 can be separated from CO_2, air, and other nitrogen oxides by using ordinary materials of construction for the chromatograph.[46] However, very strong oxidizing and fluorinating agents require the use of nickel, Monel, Teflon, and Kel-F as construction materials, and Kel-F oil on a powdered Teflon support for the column packing.[47,48] For example, a chromatograph of this sort was used for the separation of Cl_2, ClF, ClF_3, HF, and UF_6.[47] A thermal conductivity detector employing nickel[47] or Teflon-protected elements[48] has been used. The gas-density detector, where the heat filaments do not come into contact with the samples,[49,50] and flame ionization have also been used for strong fluorinating agents.

A number of metal and metalloid halides have been separated with rather conventional arrangements. For example, Keller and Freiser separated $SnCl_4$, $TiCl_4$, $NbCl_5$, and $TaCl_5$ at 200°C using a copper column packed with squalane on Chromsorb P (a modified diatomaceous earth).[51] A variety of chlorosilanes and methylchlorosilanes have been separated using silicone oil plus diethyl phthalate as the stationary-phase and thermal conductivity detectors.[52]

In his study of methyl-ethyl lead alkyls, Dawson found that pretreatment of the Chemsorb W support with sodium hydroxide reduced the interchange of alkyl groups between the lead species. SE-30 silicone rubber was used as the sta-

[40]S. A. Green, *Anal. Chem.*, 31; 480 (1959).

[41]G. Gnauck, *Z. Anal. Chem.*, 189; 124 (1962).

[42]Available from Analabs, The Foxboro Company, 80 Republic Drive, North Haven, CT 60473.

[43]L. Marvillet and J. Tranchant, in R. P. W. Scott, Ed., *Gas Chromatography, 1960* (London: Butterworth, 1960), p. 321.

[44]A. B. Littlewood, *Gas Chromatography* (New York: Academic, 1962), pp. 372ff..

[45]B. Thompson (general references).

[46]J. M. Trowell, *Anal. Chem.*, 37; 1152 (1965).

[47]A. G. Hamlin, G. Iveson, and T. R. Phillips, *Anal. Chem.*, 35; 2037 (1963).

[48]R. A. Lantheaume, *Anal. Chem.*, 36; 486 (1964).

[49]J. F. Ellis, C. W. Forrest, and P. L. Allen, *Anal. Chim. Acta*, 22; 27 (1960).

[50]T. R. Phillips and D. R. Owens in R. P. W. Scott, Ed., *Gas Chromatography, 1960* (London: Butterworth, 1960), p. 308.

[51]R. A. Keller and H. Freiser, in R. P. W. Scott, Ed., *Gas Chromatography, 1960* (London: Butterworth, 1960), p. 308.

[52]G. Fritz and G. Ksinsik, *Z. Anorg. Chem.*, 304; 241 (1960).

tionary phase and, in line with previous work, electron capture detection was found to give high sensitivity for the lead alkyls.[53]

The gas chromatographic separation of many hydrides may be accomplished with relative ease. For example, Borer and Phillips resolved 21 silanes up through Si_8H_{18} and 5 germanes through Ge_4H_{10} on a column of silicone oil on Celite.[54] This same type of column has been used to separate the boron hydrides through pentaborane,[55] while paraffin oil was used as the stationary phase in another study.[56] The borane samples of Borer, Littlewood, and Phillips included the unstable compounds B_4H_{10} and B_5H_{11}. In order to check for decomposition, they performed several repeated passes of their samples through the chromatograph and this revealed slight decomposition of B_5H_{11}. A check of this sort appears advisable whenever unstable compounds are handled. Thermal conductivity detectors have been used with all of these hydride compounds; however, less thermally stable hydrides create deposits in the detector, which impairs its function. An interesting technique which circumvents this problem and increases the sensitivity of detection involves the use of a furnace on a quartz capillary tube which is placed between the exit of the column and the detector. Complete pyrolysis of the hydride occurs in this tube to yield hydrogen, which is then detected by thermal conductivity. Highly purified nitrogen was used as the carrier gas, and the best separation of SiH_4, GeH_4, AsH_3, and PH_3 was obtained with a silicone oil stationary phase.[57]

9.3 STORAGE OF GASES AND SOLVENTS

Small quantities of gases are most conveniently stored in glass bulbs, which are selected to be free from thin spots and are well annealed. Despite these precautions, there is the chance of implosion when the bulb is evacuated; the resulting hazard of flying glass may be avoided by wrapping the bulb with tape, enclosing it in a wire mesh cage, or coating it with a tough plastic paint.[58] The requirements for a stopcock or valve on a storage bulb are much more stringent than for ordinary use because continual exposure to the gas may break down stopcock grease. A Teflon-glass valve does not suffer from this disadvantage, but when it is used on a storage bulb, the seat side of the valve should retain the gas since the stem side is more subject to leakage (Fig. 9.29).

Large quantities of gas and solvents with high vapor pressures may be stored

[53]H. J. Dawson, Jr., *Anal. Chem.*, 35; 542 (1963).

[54]K. Borer and C. S. G. Phillips, *Proc. Chem. Soc.*, 189, (1959).

[55]K. Borer, A. B. Littlewood, and C. S. G. Phillips, *J. Inorg. Nucl. Chem.*, 15; 316 (1960).

[56]J. J, Kaufman, J. E. Todd, and W. S. Koski, *Anal. Chem.*, 29; 1032 (1957).

[57]G. G. Devyatykh, A. D. Zorin, A. M. Amelchenko, S. B. Lyakhmanov, and A. E. Ezheleva, *Akad. Nauk.*, SSSR (English trans.), 156; 594 (1964).

[58]Available from Ace Glass Company, Vineland, NJ.

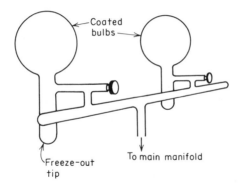

Fig. 9.29. Typical arrangement of gas storage bulbs.

in 200-mL stainless-steel cylinders fitted with packless valves and mounted firmly on an angle iron frame. The tank may be attached to a small manifold (equipped with a mercury bubbler manometer) through a glass bellows or a loop of 6-mm glass tubing to take up any slight differential movement (Fig. 9.30).

It is convenient to store frequently used solvents in containers directly attached to the vacuum line. The valve requirements are even more stringent than for gas storage, because the attack of stopcock grease by organic solvent vapors is a certainty. Therefore, Teflon-stemmed valves are generally called for. A good solvent storage container may be constructed from a distillation flask. As shown in Fig. 9.19, the solvent may be fractionally distilled directly into the sidearm of

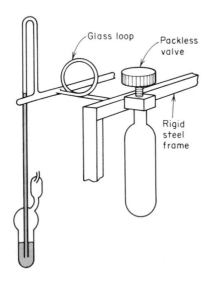

Fig. 9.30. Storage of gases in metal tanks. The valve is bolted directly to the steel frame, which is rigidly mounted.

this flask. The solvent is then frozen down, the flask is evacuated, and the side-arm is sealed off. Inclusion of a drying agent, such as calcium hydride, in the flask ensures that the solvent will remain anhydrous.

The long-term storage of liquids and gases is best accomplished in an ampule or bulb fitted with a break-seal. Once the sample is sealed off in such a container, leakage is no problem as it would be if a stopcock or valve were used. Furthermore, this practice is economically sound, since a break-seal is much less expensive than a stopcock or valve.

9.4 SEALED TUBE REACTIONS

The difficulties created by stopcocks and valves can usually be minimized. However, it is occasionally necessary to completely eliminate these sources of leakage and contamination by the use of break-seals and vacuum seal-offs. Typical situations in which sealed tube techniques are widely used are quantitative hydrolysis and oxidation reactions which require elevated pressures and temperatures, precise physical measurements on highly reactive organometallic compounds, long-term storage of reactive samples, and nonaqueous reactions under high pressure (for example, SO_2 or NH_3 at room temperature). Each piece of apparatus must be constructed to meet a specific need, so it is not possible to outline an apparatus which is of general use. Nevertheless, several examples will be presented here which serve to indicate the approach.

Break-seals (Fig. 9.31) are commonly used on sealed reaction and storage tubes because this type of seal allows recovery of volatile materials. The type illustrated in Fig. 9.31a may be opened in an apparatus which contains an off-center arm which may be rotated to break the small tube (Fig. 9.32a). Among the many variants of this design of tube opener, one by Mahler and Velmey is constructed from a 1/4-in. stainless-steel needle valve which is drilled out to receive the break-seal so that the valve stem may be screwed down on the small tube (Fig. 9.32b).[59] The break-seal and vacuum system are connected to this

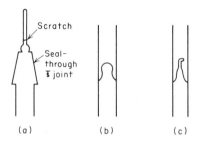

Fig. 9.31. Break-seals.

---

[59]W. Mahler and H. V. Felmey, *Rev. Sci. Instr.*, 33, 1127 (1962).

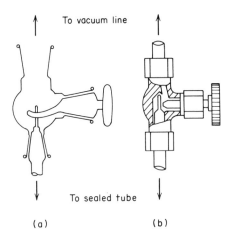

Fig. 9.32. Tube openers.

modified valve by Swagelok fittings in which a Teflon front ferrule is used (see Chapter 4). The break-seals illustrated in Fig. 9.31*b* and *c* are opened by means of a magnetic hammer which consists of a glass-encased iron rod actuated by means of a hand magnet.

The reaction tube illustrated in Fig. 9.33*a* is useful for hydrolysis reactions and other situations in which considerable pressure may build up in the tube because the break-seal used here withstands high pressures. The procedure involves distillation of the compound and outgassed water (or HCl-water mixture) into the tube on the vacuum line. It is then sealed off under high vacuum (Ap-

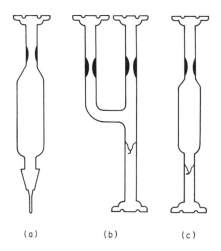

Fig. 9.33. Typical sealed tube designs.

pendix II) and allowed to warm to the appropriate temperature. When high pressures are anticipated, every precaution must be taken to avoid endangering laboratory personnel. In general, a safety-glass explosion shield or a small metal containment shield around the tube should be used, and the experiment should be arranged so that the tube can be cooled to a low temperature before it is handled. The break-seal is easily broken by clumsy handling, so it is protected by an open joint. Glass apparatus under pressure must always be treated as dangerous; however, some rough guides are given in Appendix II for the minimization of breakage. It is quite important for a glass bomb tube to be smoothly blown and well annealed.

After hydrolysis or similar reaction, the tube is cooled in liquid nitrogen to reduce the pressure of the gases and opened with a special device (Fig. 9.32). If noncondensable gases result, they are collected with the Toepler pump or low temperature adsorption for measurement and identification.

If the weight gain of the contents of the tube is of interest, the break-seal described above (Fig. 9.31a) is inconvenient because the broken parts are not easy to collect and weigh. In this case the break-seals illustrated in Fig. 9.31b and c are more useful. The former is available commercially at low cost, and either type is easily constructed. These break-seals are broken open by means of a glass-encased iron rod or, if it will not react with the gases under study, a clean ball bearing. The breaker is raised with a hand magnet and allowed to fall on the thin break-seal. The tube shown in Fig. 9.33c is useful for studying gas-solid reactions where the reaction has to be allowed to stand for a long time or where it requires heating. A typical procedure involves weighing the tube and a small stopper, introducing the solid, outgassing under high vacuum, reweighing, reevacuating, distilling a measured quantity of reactive gas back into the tube, and sealing the tube off under vacuum (Appendix II). When the reaction is complete, the tube is attached to the vacuum line, the space above the break-seal is evacuated, and, after the gas in the tube has been condensed, the tube is opened by means of the magnetic hammer. The tube is allowed to warm very slowly by means of a chilled Dewar to prevent the solid and glass particles from being carried into the vacuum system by a sudden surge of gas. After the gas is collected, dry nitrogen may be admitted to the tube, and the tube removed from the vacuum line while the magnetic hammer is quickly removed and a small rubber stopper is used to seal the open end. Both parts of the tube are then weighed in order to calculate the weight gain of the solid. In addition, the recovered gas may be measured and identified.

The trap-to-trap condensation of compounds with low volatility is always troublesome, and the situation is aggravated if the compound evolves small quantities of less condensable decomposition products. When these cases are encountered, the sample is best collected in a U-tube (Fig. 9.33b) because it allows the distillation of the compound through one leg while a high vacuum is maintained on the other. Both arms of this tube must then be sealed to isolate the sample.

Sealed-tube manipulations have been employed in physical studies on

Grignard reagents where traces of oxygen and stopcock grease may lead to erroneous results.[60,61] As illustrated in Fig. 9.34, the apparatus is designed with a specific sequence of operations in mind.

Reactions of liquefied gases, such as NH_3 and SO_2, may be carried out in sealed tubes at room temperature. Liquid ammonia exerts approximately 10 atm pressure at room temperature, so a tube diameter larger than 15 mm should be avoided and the previously mentioned safety precautions must be observed. Franklin and his students carried out many acid-base and metathesis reactions in multilegged tubes such as those illustrated in Fig. 9.35.[62] The reagents may be charged into the legs, the solvent introduced on the vacuum line, and the tube sealed off. The reaction between two or more solutions is accomplished by tipping the tube. If a solid is formed, the supernatant liquid may be decanted and the leg containing the solid may then be cooled in an ice water bath to accumulate ammonia for washing the solid. Some workers prefer to use multilegged vessels of this sort with a small pressure stopcock fitted to the inlet tube, in which

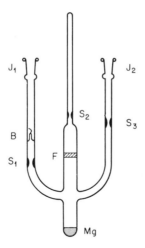

Fig. 9.34. Apparatus used for the preparation and sampling of Grignard reagents for NMR spectroscopy. The apparatus was evacuated through J_2 and the Mg baked. Excess ether and alkyl halide were distilled onto the metal, and the apparatus was sealed off at S_3. Upon warming to 0°C and shaking, a rapid reaction occurred. The apparatus was evacuated at J_1, the break-seal B was broken, and ether and excess alkyl halide were removed. Fresh dry ether and the tetramethyl silane standard were distilled in, the apparatus was sealed off at S_1, and a small portion of the solution was filtered through the frit F into the NMR tube, which was sealed off at S_2. (Adapted from D. F. Evans and J. P. Mahler, *J. Chem. Soc.*, 1962, 5125.)

[60]A. D. Vreugdenhill and C. Blomberg, *Rec. Trav. Chim.*, 82; 453 (1963).

[61]D. F. Evans and J. P. Mahler, *J. Chem. Soc.*, 1962; 5125.

[62]E. C. Franklin, *The Nitrogen System of Compounds* (New York: Reinhold, 1935).

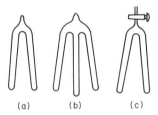

Fig. 9.35. Sealed tubes for the observation of reactions in liquefied gases. Liquid phases may be decanted from one leg to the next.

case the tube is easily opened to sample gases or remove the solvent.[63] Of course, the tube may also be fitted with a break-seal through which gaseous and volatile products may be removed. This was the approach taken by Stock, who included a slightly constricted portion in the middle of the tube with a wad of glass wool to serve as a filter.

9.5 MISCELLANEOUS TECHNIQUES

A. Sampling Alkali Metals. Of the alkali metals, lithium is in some respects the most difficult to handle. It reacts slowly with nitrogen; so when purity is important, it should be handled in an argon atmosphere or in a vacuum. In addition, the molten metal reacts with Pyrex, causing it to crack.

For rough applications it is satisfactory to cut and weigh sodium or potassium chunks under a hydrocarbon. **CAUTION: Halogenated hydrocarbons should not be used because they present a serious explosion hazard in the presence of alkali metals. In the case of potassium, the oxidized layer should be carefully scraped from the surface before the metal is cut, because an explosion may result if this crust is embedded in the metal.** For many, purposes weighed quantities of pure sodium may be introduced in sealed, thin-walled glass bulbs which may be broken with a glass-encased magnetic hammer to release the metal. These bulbs are prepared as follows. A thin glass bulb is blown onto a joint so that there is a constricted region between the bulb and the joint (Fig. 9.36). The bulb is weighed and its volume is measured for a subsequent buoyancy correction. A chunk of metal of the approximate desired weight is placed in the apparatus so it rests on the constriction, and the bulb is attached to a vacuum line and evacuated. When a high vacuum is attained, the constriction and region above it are flamed gently so that the metal melts and runs down into the bulb. The crust of oxide is left behind in this process. The bulb is sealed off under vacuum at the constriction. After the top part is cleaned, it is weighed along with the bulb, and

[63]W. C. Johnson and W. C. Fernelius, *J. Chem. Educ.*, 6; 441 (1929). In this reference and footnote 63, the procedures which are described do not involve the use of a vacuum line.

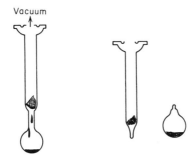

Fig. 9.36. Preparation of weighed samples of pure alkali metal. On the left the sodium is being melted under vacuum by means of a cool flame. On the right the bulb has been sealed off under vacuum.

a buoyancy correction is then applied to give an accurate weight for the metal in the bulb. Sodium, which is sealed off in this manner, is useful for quantitative work on the vacuum line since no extraneous gases or solid impurities are introduced with the sample.

The sodium-potassium alloy (45–90 weight %K) is molten at room temperature and cesium melts at 28.5°, so both of these are easily handled as liquids. For example, a hypodermic syringe is convenient for their transfer. The sodium-potassium alloy is made by heating the two metals together while they are protected by a high-molecular-weight hydrocarbon.

With the exception of lithium, the alkali metals may be distilled in an all-glass, high-vacuum apparatus to yield a very pure sample. Finally, a novel method for the quantitative introduction of small amounts of sodium into a glass apparatus is the electrolysis of the metal through glass.[64]

B. Generation of Dry Gases. Directions are given below for the preparation and/or purification of several useful gases.

Oxygen.[66] A tenfold excess of potassium permanganate (calculated on the basis of $2KMnO_4 = MnO_2 + K_2MnO_4 + O_2$) is placed in a glass bulb which is sufficiently large to allow doubling of the volume of solid as the reaction proceeds. A plug of glass wool is placed above the sample to avoid contamination of the vacuum system by solid particles. The sample is heated to 100°C while it is evacuated overnight under high vacuum. After this initial bake-out, the sample is left open to the vacuum pump while the temperature is raised to approximately 200°C, where oxygen production commences. The vacuum pump is closed off briefly so that oxygen evolution can be roughly estimated. Evacuation is continued until 5–10% of the theoretical quantity of oxygen has been dis-

[64]W. L. Jolly, *J. Phys. Chem.*, 62; 629 (1958).

carded. The oxygen sample is then collected while the solid is heated above 200°C (but not in excess of 230°C).

Methane.   Research Grade (about 99.65 mole % pure) and CP (about 99.0 mole % pure) methane are available in cylinders. The removal of CO_2, O_2, N_2, and higher hydrocarbons from methane by adsorption on charcoal has been described in the literature.[65]

Ethylene.   This gas is available from the Matheson Company in CP (99.0% pure) and Research Grade (99.9 % pure). Oxygen may be removed by passing the gas through a tube containing MnO.

Sulfur Dioxide.   High-purity sulfur dioxide is available in cylinders. It may be further purified by passing it over P_4O_{10} to remove water and degassing. In their precise work on the vapor pressure of SO_2, Giauque and Stephenson prepared this gas by the reaction of concentrated sulfuric acid with sodium sulfite.[66] It was then passed through water to remove SO_3, through P_4O_{10} to remove the water, and finally subjected to several trap-to-trap distillations, following by degassing.

Ammonia.   The major contaminants in the "anhydrous ammonia" available from the Matheson Company in cylinders are water, oil, and noncondensable gases. Most of these impurities are removed by passing the NH_3 through a trap held at $-22°C$ and condensing it at $-196°C$ under vacuum. The water is removed by distilling the ammonia into a tube which contains a small lump of sodium. The tube is warmed slightly to melt the ammonia and form a blue solution, and the resulting dry ammonia is pumped away. A fairly large plug of glass wool must be included between this tube and the vacuum system to avoid contaminating the line with a fine spray of sodium metal. The gas is condensed at $-196°C$ under high vacuum to remove hydrogen and other noncondensables.

GENERAL REFERENCES

Temperature and Temperature Measurement

Benedict, R. P., 1984, *Fundamentals of Temperature, Pressure, and Flow Measurements*, 3rd ed., Wiley, New York.

Quinn, T. J., 1983, *Temperature*, Academic, London. A comprehensive review of thermometry over the temperature range 0.5–3000 K. The book includes chapters on resistance thermometry and thermocouples, and is extensively referenced.

[65]J. H. Eiseman and E. A. Potter, *J. Res. Nat. Bur. Std.*, 58; 213 (1957); H. H. Storch and P. L. Golden, *J. Am. Chem. Soc.*, 54, 4662 (1932).

[66]W. F. Giauque and C. C. Stephensen, *J. Am. Chem. Soc.*, 60; 1389 (1938).

Schooley, J. F., Ed., 1982, *Temperature—Its Measurement and Control in Science and Industry*, Vol. 5, American Institute of Physics, New York. An excellent source of state-of-the-art thermometry comprised of papers from the Sixth International Temperature Symposium. Topics which are covered include temperature scales and fixed points, radiation, resistance, thermocouple, and electronic thermometry, temperature control, and calibration techniques. Preceding volumes in the series date back to 1939.

Gas Chromatography

Keszthelyi, C. P., 1984, in P. T. Kissinger and W. R. Heineman, Eds., *Laboratory Techniques in Electroanalytical Chemistry*, Marcel Dekker, New York. Vacuum line design and experimental methods for electrochemical experimentation.

McNair, H. M., and E. J. Bonelli, 1969, *Basic Gas Chromatography*, 5th ed., Varian Associates, Inc.

Thompson, B., 1977, *Fundamentals of Gas Analysis by Gas Chromatography*, Varian Associates, Inc., Palo Alto, CA.

10

Metal Systems

Glass apparatus often is widely used for the construction of laboratory apparatus because it is relatively inert, light, transparent, and easy to clean. However, metal apparatus has decided advantages in some applications. For example, metals are mechanically more robust than glass, many metals are resistant to reactive fluorides, and metal valves and joints can be much more leak-tight than their glass counterparts. Because of these qualities, metal is used for the construction of high-pressure apparatus, fluorine-handling vacuum lines, and very-high-vacuum systems. In addition, metal is a popular material for gas-handling manifolds and flow reactors for catalysis research.

The first part of this chapter is devoted to the proper handling of gas cylinders and compressed gases. This section should be required reading for anyone working with compressed gases for the first time. The construction of apparatus from seamless tubing is described next, and the design and characteristics of common metal components, such as valves and joints, are presented. The chapter ends with details of some general-purpose metal apparatus for catalytic reactors and fluorine chemistry. Attention also is directed to Appendix IV for details on the properties of metals.

10.1 COMPRESSED GASES

A. Safe Handling of Cylinders. Several popular cylinders for the containment of compressed gases range from the size often used for nitrogen or oxygen (about 9 in. in diameter by 52 in. high, exclusive of the valve and cap) to the lecture bottle (2 in. in diameter by 15 in. long) which is used to dispense small amounts of reactive gases. In the United States, the cylinder sizes vary from one supplier to the next; however, there is fairly uniform adoption of valve outlet fittings as perscribed by the Compressed Gas Association. A typical compressed gas valve and fitting for large cylinders consists of an emergency pressure release fixture, an on-off valve, and the outlet fitting. **CAUTION: Never tamper with the emergency pressure release fixture. Unfortunately, the emergency pressure release fitting sometimes has a hexagonal outer contour and therefore can be loosened by means of a wrench.** The valve assembly is the most vulnerable part of the cylinder and it can be broken off if a cylinder is accidentally dropped. A pressurized cylinder which has broken open is very dangerous because it is propelled like a rocket. **CAUTION: To minimize the chance of a cylinder being ruptured by falling, the cylinders should always be strapped in place when the protective cap is removed, and the cap should be replaced before the cylinder is moved to another location.** Of course, a cylinder should never be deliberately dropped, even when the protective cap is in place.

Some of the common high-pressure cylinder fittings are illustrated in Fig. 10.1. Many of these have as their sealing surfaces a cone in the outlet and a cone-shaped or rounded nipple on the connecting piece. Those fittings, which do not have gaskets, require firm tightening to achieve a gastight seal. For a few gases which are delivered at lower pressure, the connection is made by means of flat surfaces, one of which contains a washer of lead or other soft metal. **NOTE: Left-handed threads are used on flammable gas cylinder fittings, for example, H_2, propane, and CO. The connector for these flammable gases can be identified by means of V-shaped indentations on the edges of the hexagonal nut (Fig. 10.1.b).** It may seem to the user that the wide variety of outlets is a nuisance; however, they serve the beneficial purpose of avoiding accidental connection of the wrong gas to an apparatus, and they avoid the possibility of using a regulator for a flammable gas on a cylinder containing oxygen or another oxidizing agent. Trapped gas or oil residues from a flammable gas are likely to explode if suddenly compressed with high-pressure oxygen.

The Compressed Gas Association specifies two types of outlets on lecture bottles (Fig. 10.2), depending on whether or not the gas is corrosive. A lead washer is inserted between the tank outlet and the connector fitting. These gaskets generally last for only a few cycles of assembly and disassembly, so a good stock should be kept on hand. These are readily made using two hole punches of appropriate diameter and a sheet of $1/16$-in.-thick lead. As with the larger tanks, the outlet valve on the standard lecture bottle is of the on-off type and is not suitable for the control of gas. Both pressure regulators and needle valves are available which will mate with the lecture bottle fitting. A needle valve is the most generally useful means of gas regulation because lecture bottles are most often used to dispense small quantities of gas for chemical reactions. In place of the lecture bottle, some chemical supply houses provide gases and highly volatile liquids in small welded steel containers (about 3 in. in diameter and 5 in. high) that are light and easy to handle. In general, these cylinders are used for low pressure and relatively noncorrosive gases, and are fitted with a needle valve outlet equipped with a swage or pipe thread fitting. The valve that is supplied on the

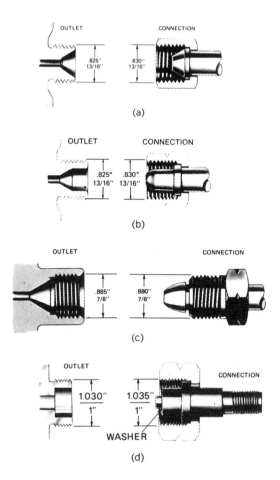

Fig. 10.1. Some common high-pressure tank fittings. The seal in the top three fittings is made by contact between the nipple in the connection and the seat in the outlet. Note the cone-shaped, round, and bullet-shaped nipples for (a), (b), and (c) respectively. The bottom connection (d) is sealed by means of a flat soft-metal gasket. Never attempt to mix outlet and connections between the various designs. Note the grooves on the connection nuts for connections (b), (c), and (d). This groove indicates a fitting with left-handed threads. (Reproduced by permission of the copyright holder, Matheson Gas Products, Inc.)

tank often is satisfactory for controlling the gas outlet, but if there is any doubt, an additional needle valve (see Section 10.5) or pressure regulator should be added.

B. Safe Delivery of Gases. The vast majority of compressed gas cylinders are supplied with efficient on-off valves which provide no significant flow or pressure control. **CAUTION: It is necessary to equip a compressed gas cylinder with some means of controlling the pressure and/or flow of the gas being deliv-**

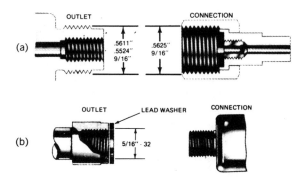

Fig. 10.2. Fittings for lecture bottles. Both fittings are designed to seal by means of a soft lead washer, which must be replaced frequently. Fitting (*a*) is for noncorrosive gases and (*b*) is for corrosive gasses. (Reproduced by permission of the copyright holder, Matheson Gas Products, Inc.)

ered. The choice of pressure versus flow control is based on the nature of the experiment being performed. The delivery of gas to a closed, low-pressure system requires the use of pressure control by means of a diaphragm regulator (Fig. 7.14). For example, the delivery of nitrogen or argon to an inert atmosphere Schlenk-type apparatus (Chapter 1) dictates the use of a good pressure regulator which will deliver about 3 psig of gas. Similarly, the oxygen for a glassblowing torch is supplied via a diaphragm regulator. The least expensive pressure regulators are fitted with only one stage of regulation. This results in a slow change in the outlet pressure as the tank pressure decreases. Much better regulation is obtained with two-stage regulators. Another point which requires attention in the choice of a regulator is the output regulation range. A different regulator is chosen for high-pressure outlets (to fill high-pressure reactors) than for general use. The general-use models listed by many distributors have output regulation in the 0–75 or 0–50 lb/in.^{2} range, which is not optimum for some of the most common laboratory applications. The best choice for inert-gas lines and glassblowing torches is a regulator designed for a 0–15 lb/in.^{2} output. As described below, corrosive and high purity gases require special regulators.

If gas is to be delivered to a reaction flask which has an unobstructed outlet, a simple flow control valve on the high-pressure cylinder will provide adequate regulation of the gas delivery. In this case a needle valve is attached to the cylinder, or to a pressure regulator which in turn is attached to the cylinder. It also is possible to deliver gas to a closed system, such as a vacuum line, with a flow control valve. In this case the pressure within the apparatus must be carefully monitored by means of a manometer and the system should also be equipped with a means of pressure relief, such as a mercury bubbler manometer (Fig. 7.2).

When reactive gases such as HCl or BF_3 are handled, the regulator or valve should be constructed of a metal which is corrosion resistant. The gas distributor's catalog generally lists the appropriate valve material.[1] For example, a needle valve controller for the delivery of hydrogen halides or metalloid halides from a lecture bottle should be constructed from Monel, whereas brass valves will suffice for unreactive gases such as Freons, sulfur hexafluoride, and the like. The combination of mechanical action of the valve, the corrosive action of the gas, and moisture from the air can readily lead to corrosion and seizure of valves or regulators. Therefore, it is good practice to purge out the valve after each use. In the case of a lecture bottle needle valve, this entails removal of the needle valve and flushing it with a stream of inert gas. Some of the commercially available regulators and fittings for corrosive gases are equipped with a purging inlet so the system can be purged with inert gas before and after the reactive gas is dispensed.

When very-high-purity gases are handled, special pressure regulators with metal diaphragms are desirable to avoid contamination by atmospheric gas diffusion through elastomeric materials on conventional regulators. Similarly, metal or glass delivery lines are preferred over rubber for the delivery of high-purity gas.

10.2 CUTTING AND BENDING TUBING

Seamless copper and stainless-steel tubing are used for a variety of gas delivery and reactor systems in the laboratory, and it is a tremendous convenience and time saver if the research worker is able to perform simple fabrication and repairs. Seamless tubing of 1/4-in. O.D. or larger is fairly rigid; therefore, when high flexibility is needed, a loop of 1/8-in. seamless stainless-steel tubing may be used. An even better choice for low-pressure applications which require flexible metal tubing is thin-walled corrugated stainless-steel tubing. The methods of joining corrugated tubing to apparatus will be described in this section.

Seamless copper or stainless-steel tubing may be cut with a hacksaw or, more conveniently, with a tubing cutter by the process illustrated in Fig. 10.3. In either case, the cut should be square and clean if fittings are to be used on the tube ends. Any burr on the tube can be removed with a file or the special routing blade attached to most tubing cutters. When bent on a sharp radius, tubing tends to collapse. To reduce this problem, a hand-operated tubing bender is very helpful (Fig. 10.3). Sturdy, tightly coiled spring sets may be purchased for bending tubing. The spring is slipped over the outside of the tube and removed after the bend is completed. It is also possible to minimize kinking by packing the tube with sand and corking the ends. The advantage of these latter two methods are that the bend can be made with any radius around the circumference of a pipe or bar of suitable size. If the exact radius is not critical, the

[1]*Airco Rare and Specialty Gas Catalog*, Airco, 575 Mountain Ave., Murray Hill, NJ 07974; *Linde Specialty Gases and Related Products*, Union Carbide Corp., Linde Div., Box 444, Somerset, NJ 08873; *Matheson Gas Products*, Matheson Gas Products, P.O. Box 1587, Secaucus, NJ 07094.

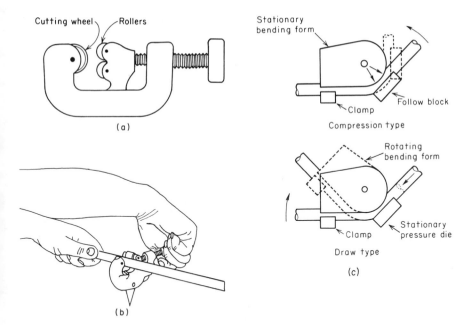

Fig. 10.3. Cutting and bending tubing. (*a*) Side view of a tubing cutter. (*b*) The tubing cutter in use: The handle is turned until the cutting wheel slightly penetrates the metal tubing. (*b*) The cutter is then rotated around the tubing. The handle is retightened and the process is repeated until the tubing is completely severed. (*c*) Two designs for tubing benders. (Reproduced by permission of the copyright holder, Crawford Fitting Co., Cleveland, Ohio.)

tubing bender is generally preferred. Large-diameter tubing is not easily bent, and sharp turns are impractical in any diameter tubing, so in these cases an elbow fitting is used.

Thin-walled corrugated tubing can be crushed by swage-type fittings. Sometimes this type of tubing is fitted by the supplier with heavy-walled end tubes which can be connected by conventional techniques (Section 10.3) to an apparatus. If the tubing is not so equipped, it is necessary to have thick-walled tubing brazed or welded to the ends. One manufacturer also provides heavy-walled inserts which can be slipped inside the thin-walled tubing to provide added rigidity.[2]

10.3 TUBING, JOINTS, AND FITTINGS

A. Solder and Weld Fittings. Copper elbows, T's, and crosses are manufactured to be soldered or brazed to copper tubing (Fig. 10.4). When soft solder

[2]Crawford Fitting Co., 29500 Solon Rd., Solon, OH 44139.

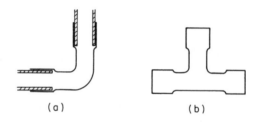

Fig. 10.4. Two solder fittings. (*a*) Cross-section of an elbow. (*b*) A T-fitting. (Reproduced by permission of the copyright holder, Crawford Fitting Co., Cleveland, Ohio.)

is used, the inside of the fitting and outside of the tubing are tinned with solder and the hot parts are slipped together. An acid-type flux, for example, 10% HCl plus 10% $ZnCl_2$ in water, aids in the tinning process. Similarly, hard solder (brazing or silver solder) joints can be used. Soldered joints are compact and can be very leak-tight, but they are permanent and the soldering process introduces impurities and/or oxidation of the tubing; therefore, the apparatus should be thoroughly cleaned before it is used. Weld fittings, similar in general design to the solder fittings, are available in a variety of materials.[2]

B. Flare Fittings.   This type of fitting (Fig. 10.5) is one of the oldest means of making a removable tubing joint. As shown in the figure, annealed tubing with a smoothly cut end is flared by clamping it into a flaring tool and tapping the inner die on this tool with a mallet. If the metal is not annealed or if the flaring process is roughly performed, a crack may develop on the flared edge, rendering that surface useless. The flared tubing forms a joint between a nut and a cone of the flared tubing. Considerable torque is needed to obtain a leak-tight seal. The flare fitting is cheap and works well with copper, but it is more trouble to make than swage-type fittings and is not practical with many hard metals. To overcome these disadvantages, a variety of swage-type fittings have been developed under trade names such as Swagelok, Gyrolok, and Tylok.

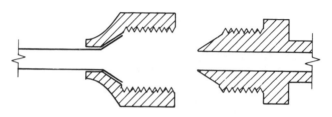

Fig. 10.5.   Cross-section of a flared tubing fitting. Annealed tubing is flared by means of a simple flaring tool, and this flared tubing is compressed against a nipple fitting by means of a threaded nut.

C. Swage-Type Fittings. These popular connectors are available in various metals and plastics, and they are useful for joining similar or dissimilar materials, such as metal to plastic or metal to glass.[2,3] The common swage tubing coupler has four parts: the body, into which the tube is placed, two ferrules, and a packing nut (Fig. 10.6). In general, the tubing should be softer than the fitting so that during the tightening process the ferrules sink into the tubing and at the same time the end of the tubing is butted tightly into the body. In the case of stainless steel, hardened tubing should be avoided because it is not sufficiently ductile to conform to the ferrules and body. In general, brass fittings are used with copper tubing, stainless-steel fittings with stainless-steel tubing, Monel with Monel, and so on. The joint is assembled as illustrated in Fig. 10.6 and the nut turned down finger tight. By means of a wrench on the body and another on the nut, further tightening by $1\frac{1}{4}$ revolutions completes the assembly. These types of fittings can be opened and closed; however, the ferrules are not removable. When a fitting is reassembled, a fraction of a revolution is sufficient to complete the seal. The swage-type fittings begin to leak if opened numerous times, and the flare joint described above or the modified flange-type joints described below are preferred when frequent opening is necessary.

Swage-type joints also are convenient for joining metal apparatus to glass apparatus. In this case, the body of the joint may be made of any metal and a Teflon or graphite front ferrule is employed. The ferrule deforms on compression to make the seal. Because Teflon will cold-flow, it is sometimes necessary to retighten these joints. A graphite front ferrule is preferred when elevated temperatures are employed, as in the case of heated gas chromatography columns. In either case, the backing ferrule is generally chosen as a harder material than the front ferrule, for example, nylon or metal in the case of a Teflon front fer-

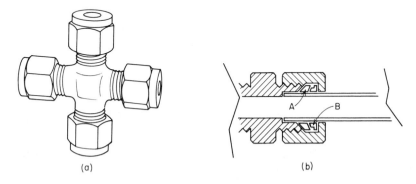

(a)   (b)

Fig. 10.6. Swage-type fittings. (*a*) T-fitting (*b*) Cross-section showing the front ferrule A and the back ferrule B. (Reproduced by permission of the copyright holder, F. J. Calahan, "Swagelok Manual," Crawford Fitting Co., Cleveland, Ohio.)

[3]Hoke Inc., 1 Teanakill Park, Creskill, NJ 07626.

rule, or metal in the case of a graphite front ferrule. It is critical to employ tubing which makes a precise fit with the Swage fitting and ferrules. Unfortunately, most swage-type joints available in the United States are made to English standards, whereas the standard-walled borosilicate glass tubing is manufactured to metric specifications. Because the tolerances on ordinary glass tubing are not highly precise, it is possible to find the occasional piece of standard-walled tubing which will match a swage fitting with English dimensions. The better solution is to employ medium-walled tubing, which is manufactured to English dimensions. It is important to have a square cut and no sharp edges on the end of the glass tube, because sharp or jagged edges will cut or scratch the soft ferrule materials and irregular edges are likely to break. It is possible to achieve a satisfactory edge by means of a clean break and a very light fire polish, but the preferred method is to cut the tube end on a diamond saw and remove the edge by very light grinding on a water-cooled sanding belt or similar grinding surface.

Most gas chromatography supply houses also stock one-piece ferrules made of Teflon, graphite, or Vespel (a polyimide). The upper working temperature quoted for each of these is 250, 450, and 350°C, respectively, and all of these are quite useful for connecting glass tubing to metal fittings. The Vespel and graphite ferrules have the advantage over Teflon that they do not cold-flow and they give generally better performance on joints that are heated and cooled. Vespel is more mechanically robust than graphite. New ferrules made from a Vespel-graphite composite can be used up to around 400° and have even better mechanical properties than Vespel.

D. Flange Tubing Connectors.   Recently, metal tubing connectors have become available in which inner and outer packing nuts are employed to hold two flanges together.[2] In some designs an O-ring seal is made between these two flanges, and in others a metal gasket is employed (Fig. 10.7). These flange fittings are attached to tubing by soldering or welding, so they are less convenient to install initially than the swage-type fittings, but they are easily assembled and disassembled and are preferable when the system is opened often. Since the ends of the fitting are flat, these types of fittings can be used where a component must be removed that is in line with the entrance and exit tubes. This is not possible with standard swage or flared connectors. The Cajon VCR fitting, which is available in stainless steel or Monel, gives good service in fluorine vacuum systems. In this application, flat gaskets of copper or nickel can be used.

E. Miscellaneous Tubing Connectors.   The variety of tubing couplers is enormous and a comprehensive list would be out of place in this book, but several additional types will be mentioned here which are useful in the laboratory. Metal standard taper and metal ball joints are available, and these will mate with corresponding ground-glass joints to provide one means of connecting parts made of dissimilar materials.[4] As with glass standard taper joints, the seal is

[4]Kontes Glass Co., P.O. Box 729, Vineland, NJ 08360.

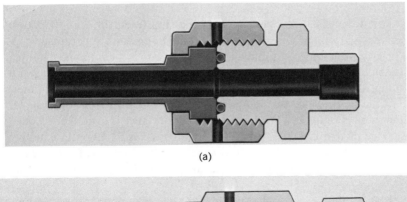

(a)

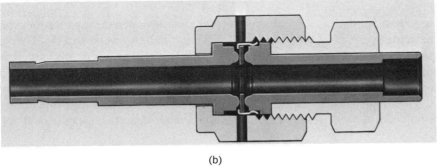

(b)

Fig. 10.7. Flange tubing connectors. (*a*) Cajon VCO O-ring sealed fitting; (*b*) Cajon VCR metal gasketed flange fitting. (Reproduced by permission of the copyright holder, Markad Services Co.)

made with grease or wax. Another useful type of joint is based on an O-ring which is compressed between the inside of the joint and the outer diameter of a tube. Plastic annular O-ring fittings of this type, designed primarily for glass apparatus, are described in Section 8.1.B, Fig. 8.5. Similarly, an annular O-ring joint, which was originally designed for a metal apparatus, is illustrated in Fig. 8.5. This joint is equally serviceable with plastic or glass tubing. As with the swage-type connectors, glass tubing must be chosen with an outside diameter which conforms to the dimensions of the fitting. This generally dictates the use of medium-walled tubing with fittings made to English dimensions. Also, the ends of the glass tubing must be prepared as described for the swage joint, so that the O-ring is not damaged. These annular O-ring seals are generally not satisfactory for pressures above 1 atm. One useful feature of this type of joint is that the tube inside the fitting may be rotated without breaking the seal.

10.4 HEAVY-WALLED TUBING AND PIPE JOINTS

Many of the fittings described above may be employed with heavy-walled tubing or metal pipe. Some fittings which are more specialized for this purpose are described here.

A. Cone Joints. The cone-type fitting, illustrated in Fig. 10.8, was designed for used with heavy-walled tubing for high-pressure apparatus.[5] A special tool or lathe is employed to shape the tube end into a cone. In one modification the outside of the tube is threaded to receive a small collar. When the gland nut is tightened against this collar, the cone is forced against the conical inner surface of the body. Since the two parts have a slightly different cone angle, the contact area is small and good leak-tight performance is achieved. At one time these cone joints were often used on vacuum systems for fluorine handling because the joint can be made and broken repeatedly. However, flange joints have generally superseded the cone joints in this applicaton.

B. Pipe Joints. Good joints suitable for vacuum or pressure applications can be made by means of tapered pipe thread joints in heavy-walled tubing. Unlike all of the joints discussed so far, the seal is made in the threads of this joint. The threaded portions of both the male and female pipe joint are tapered, and as a result the inner and outer threads are wedged into each other when they are screwed together (Fig. 10.9). To achieve a good seal the male pipe threads are wrapped with two layers of Teflon pipe tape, which serves the role of sealant and

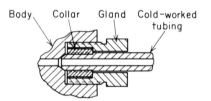

Fig. 10.8. Cone high-pressure joint. In this particular design the heavy-walled tubing is beveled on the end and threaded to receive a collar. The gland nut thrusts against the collar and forces the cone-ended tube into the body.

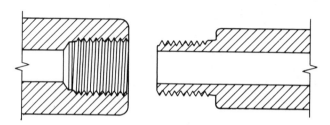

Fig. 10.9. Normal pipe threaded joint. Note the tapered threads on both the male and female components. Unlike most other joints, the seal is formed in the threaded portion of the joint.

[5]Autoclave Engineers, Inc., Erie, PA.

a lubricant for the threads. **NOTE: The use of Teflon tape should be confined to pipe threads. It makes no sense to use Teflon tape on the threads of flared, flange, or swage-type fittings because the seal is not made at the threaded portion of those fittings.**

The commonly used tapered pipe thread in the United States is designated as Normal Pipe Thread (NPT), whereas the design specified by the International Standards Organization is designated as an ISO tapered pipe thread. Tapered pipe-threaded joints can be opened occasionally. If frequent opening is necessary, other joints, such as the ISO parallel-threaded pipe joint, are preferred. When pipe-threaded joints are specified without any further qualifier in the United States, it is generally understood that the fitting is NPT. The ISO parallel pipe thread is based on parallel threads and an O-ring or other gasket butt-type seal. **NOTE: Neither ISO fittings nor standard threaded fittings should be screwed into tapered pipe threads.**

C. Bolted Flange Closures.   The flange joint is widely used to link large-diameter tubing and to provide access to the interior of reaction or vacuum vessels (Fig. 10.10). Typically, the flanges are bolted together and are sturdy enough to withstand this pressure without appreciable distortion. The bolts are spaced evenly around the flange to distribute the load evenly on the gasket material. Very often one of the flanges is cut with a groove to take an O-ring or sometimes a metal V-ring. The joint is designed so that there is significant but limited compression on the O-ring. The limit is achieved by metal-to-metal contact of the flanges. The groove is larger than the width of the O-ring, to permit radial expansion of the O-ring when it is under compression. The general strategy in O-

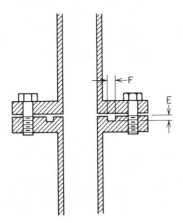

Fig. 10.10.   O-Ring sealed bolted flange closure. This type of joint is used for large diameter openings. The groove to receive an O-ring seal is machined so that the depth E is 80% of the O-ring width and F is 135% of the O-ring width. The extra clearance is on the inside of the groove for applications in which the apparatus is to be pressurized and on the outside of the groove when the apparatus is to be evacuated.

ring groove design is to specify the depth of groove to be 80% of the width of the O-ring and the width of the groove to be 135% of the O-ring width. As illustrated in Fig. 10.10, the design of the groove differs for vacuum and pressure applications.

For ultrahigh-vacuum and many high-pressure applications, O-rings do not provide the required degree of seal, so metal gaskets are employed (Fig 10.11). Annealed copper or soft aluminum gaskets are the most common, and the flanges are machined with opposing circular ridges or steps which provide a small area of contact with the gasket and thereby a high force per unit area. Flanges of this type are commercially available.[6] To maintain alignment of the two halves, guide pins are sometimes incorporated into one flange with holes in the opposing part.

10.5 VALVES

Because they involve moving parts, valves are generally more troublesome in a vacuum or pressure system than are joints. Furthermore, valves are readily damaged if improperly handled and they are often designed for a specific purpose. Because of these factors, the intelligent use of valves requires a knowledge of their construction and mode of operation. For example, a metering valve and a cutoff valve have different internal design, although they may look identical from the outside. Similarly, the nature of the seat material and the way in which the seal is made around the valve stem are often essential information in the choice of the proper valve and the maintenance of a metal system.

A. Valve Seats and Stems. Valves which are intended for fine gas flow control are designed with a long needlelike stem and a gently tapered seat (Fig. 10.12a). Although these "metering valves" give good flow control, they do not in general give a positive shutoff. When the valve is designed for positive shutoff a

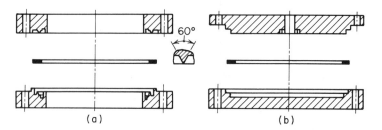

Fig. 10.11. Metal gasket sealed bolted flange closures. A soft metal gasket, generally made of aluminum or copper, is clamped between knife edges (*a*) or stepped flanges (*b*).

[6]Varian Vacuum Accessories, Varian Associates, Palo Alto, CA.

blunt stem tip and simple seat is employed (Fig. 10.12b). Cutoff valves are available for which some degree of flow control is achieved by means of a fairly sharply tapered conical tip. However, if good flow control and positive shutoff are required, a metering valve and a cutoff valve in series are generally called for. This type of configuration has already been discussed in connection with high pressure cylinders. One of the most common errors in the use of small valves is to shut them too tightly. A good shutoff valve is designed so that there is fairly large force per unit area between the stem and seat; therefore, if the valve is overtightened, the seat and/or stem is distorted and the valve will not seat properly in subsequent use. Similarly, metering valves are often delicate and not designed to be overtightened. In some valve designs, wear on the seat and stem tip is minimized by means of a "floating tip" which permits the valve stem to rotate independently of the tip. Thus, as the valve is tightened, the tip is not ground into the seat.

B. Packed Valves.   Some of the least expensive and most common valves have a sealing material (packing) compressed between the valve stem and the valve body. For valves used in laboratory apparatus this seal is generally accomplished by means of an O-ring (Fig. 10.12a), or a Teflon washer. Compression on the O-ring is determined by the groove dimensions on the stem and inside diameter of the valve body. If a leak develops along the stem, the valve must be disassembled, cleaned, and the O-ring replaced. The situation is different with a Teflon packing washer. In this case the packing is caused to press against the stem and the valve body by means of the packing nut (Fig 10.13). This packing nut may require occasional tightening to maintain proper contact with the valve stem and body. Clearly, it is important to identify the valve type in order to remedy a leak past the stem. In dealing with O-ring packed valves, it also is important to recognize the chemical sensitivity of the O-ring and, if necessary, to change the O-ring to match the compounds being handled. (See Chapter 5 and Appendix III for information on the solvent sensitivity of various elastomers.)

Another variation in the design of packed valves is the relative position of the threads which actuate the valve stem and the packing material. With the O-ring valves it is common to position the threads so they are protected from liquids or gases inside the valve by the O-ring seal (Fig 10.12a). However, with Teflon and similar packing materials, which must be maintained under pressure by the packing nut, it is more common for manufacturers to thread the valve stem inside of the packing material. This arrangement is undesirable when highly corrosive liquids are being handled, because exposure of the threaded surfaces to the corrosive material is likely to seize up the valve. Some Teflon-packed valves are available in which the threads are protected by the packing (Fig. 10.13).

C. Diaphragm and Bellows Valves.   For high-vacuum applications, the leakage of a packed valve is sometimes unacceptable, so diaphragm or bellows valves are employed. In these valves the packing is eliminated by using a flexible metal diaphragm or bellows sealed to the valve stem and the body (Fig.

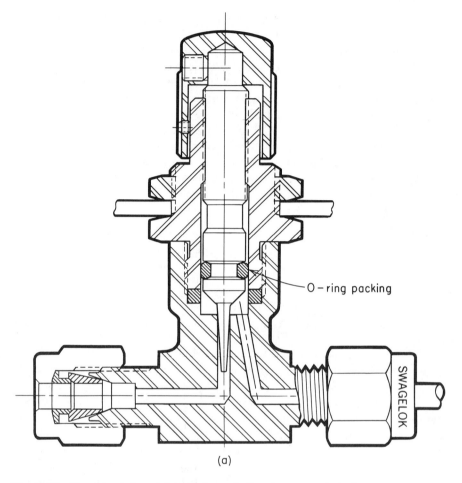

(a)

Fig. 10.12. Two valve designs. (*a*) An O-ring seals the valve stem to the body.

10.12b). The diaphragm or bellows flex as the stem is actuated and thus permit the transmission of the throttling action of the valve without the use of packing. A variety of designs are possible; in some valves the diaphragms and valve stems may be replaced because the diaphragm or bellows is attached to the valve body by means of a demountable seal involving a gasket or O-ring. In valves designed for very-high-vacuum performance, the diaphragm or bellows is sealed to the body by brazing or welding.

D.   Ball, Plug, and Gate Valves.   The valves discussed above often introduce considerable restriction into the system because of small orifices and tortuous flow patterns. Such restrictions are not serious for most applications at 1 atm pressure or above, but they may be unacceptable for vacuum applications, where small-diameter openings are a serious impediment (see Chapter 6). Ball and

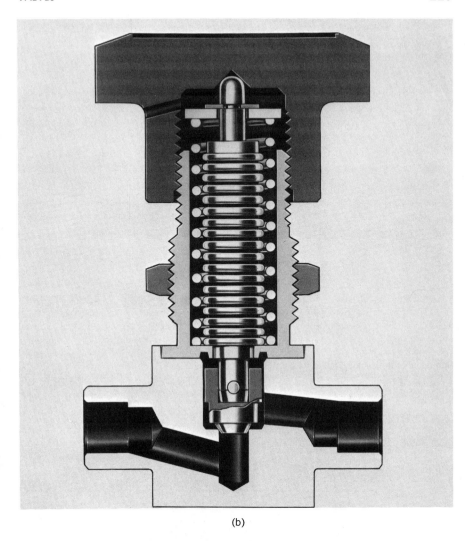

(b)

Fig. 10.12(b). A metal bellows attached at one end to the body and at the other end to the stem forms a highly leak-tight but bulkier and more expensive seal. (Reproduced by permission of the copyright holder, Markad Services Co.)

plug valves, which provide large orifices and straight lines of flow (Fig. 10.14), bear a strong similarity to glass stopcocks.[2] In the ball valve, a smooth ball, with a large hole along one axis and a handle attached along an orthogonal axis, is surrounded by a spherical cavity and an elastomer seal. A 90° rotation of the valve body will close or open the valve. The plug valve is similar, but in this case a cylindrical plug is used and the plug is equipped with O-ring seals. Both types of valves can be manufactured with close tolerances and are suitable for medium- and sometimes high-vacuum applications. The more expensive gate valve is

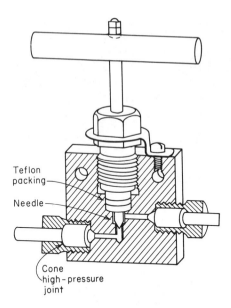

Teflon packing

Needle

Cone
high-pressure
joint

Fig. 10.13. A high pressure valve with Teflon packing. The threads on the valve stem (not shown) are above the Teflon packing and are thus protected from corrosive materials. The packing nut tightens the Teflon packing washer against the needle valve stem and the valve body. Occasional tightening of the packing nut is required to maintain a good seal.

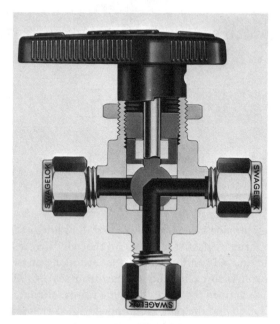

Fig. 10.14. A metal ball valve. This type of valve shown here in a 3-way version, provides a large unobstructed aperature. (Reproduced by permission of the copyright holder, Markad Services Co.)

available with a bellows closure along the stem. This type of valve is often used between a diffusion pump and an associated metal high-vacuum system.

E. Mounting Valves. It is often desirable to mount a valve securely, so that when it is actuated there is no mechanical stress on the connection between the valve and the tubing. For example, if a valve is attached to small-diameter tubes by means of swage-type joints, repeated flexing of these joints can open up leaks. Occasionally, mounting holes on the valve body are provided by the manufacturer for bolting the valve to apparatus. The second common fixture is a bulkhead connector, which consists of threads on the outside of the valve body and a pair of nuts on this threaded section. To use this mount, the valve handle is generally temporarily removed and the valve is mounted by means of these nuts through a hole in a sturdy section of sheet metal, as illustrated in Fig. 10.12b. Gate valves are generally equipped with bolted flange conectors so that they can be connected to large-orifice tubing.

10.6 PRESSURE GAUGES

The measurement of pressure from a few torr to 10^4 torr in an all-metal system can be accomplished by means of a metal Bourdon or diaphragm gauge with a dial readout. Although a bronze chamber is the most common, these gauges are available with more corrosion-resistant metals, and some are available which are accurate to 0.5 torr.[7]

Various types of pressure transducers with digital readout are available with metal interior construction. These have become quite popular for fluorine-handling systems and the like. A description of some of the common types is given in Chapter 7.

10.7 TYPICAL METAL SYSTEMS

A. Fluorine Handling Systems. Fluorine is thermodynamically capable of reacting with most materials of construction, but the reaction rate may be extremely slow below certain critical temperatures. The slow rate of corrosion of many metals occurs because of the formation of a passive fluoride film. Some generalizations on the resistance of various materials to fluorine are given by Peacock.[8] Fluorine which is free from hydrogen fluoride may be handled in borosilicate glass for periods of up to a few hours and temperatures up to 200°C without extensive attack; fused silica is suitable for longer periods and slightly higher temperatures. However, surface moisture on the glass, which is extremely

[7]American Chain and Cable Co., Helicoid Gauge Div., 230 Park Ave., New York, NY.

[8]R. D. Peacock, in *Advances in Fluorine Chemistry*, M. Stacey, J. C. Tatlow, and A. G. Sharpe, Eds., Vol 4, (Washington DC: Butterworth, 1965), p. 31.

hard to eliminate, catalyzes the reaction of fluorine with these glasses, and therefore, glass is rarely used for fluorine handling. Copper may be used up to 500°C, while nickel and Monel are useful to about 800°C. Fluorite and ceramic alumina are resistant to 1,000°C, but the latter becomes porous even though it is not eroded by fluorine. Stainless steel and brass have been used occasionally to contain fluorine at room temperature, but they are less desirable than Monel, nickel, or copper for laboratory apparatus. Passivated aluminum is satisfactory for fluorine handling even at temperatures somewhat above room temperature, but it has the practical disadvantage of not being readily soldered. Mixtures of fluorine and hydrogen fluoride are more reactive than either component, and a fair number of fluorine derivatives, such as ClF_3, are more reactive than F_2. For these highly reactive fluorides, nickel or Monel are the preferred metals of construction.

The strongest and least porous joints are made by inert-gas welding (see Appendix IV), but serviceable silver solder joints are possible. Lead gaskets, soft solder joints, and brazed joints may be used with fluorine and most all gaseous fluorinating agents, but are unsatisfactory with liquid BrF_3. Soft solder (lead-tin solder) is very sensitive to HF. Junctions between dissimilar metals, such as copper and nickel, are subject to electrolytic corrosion when exposed to liquid fluorinating agents which also are electrolytes, such as BrF_3 and F_2-HF.

Most metal systems are passivated by conditioning with fluorine and/or volatile fluorides prior to use. It is often desirable to precede the conditioning step with a hydrogen reduction of the interior of the apparatus because the mixed oxyfluoride of an element, which might be produced from oxide coatings in the apparatus, are frequently similar in volatility to the simple binary fluorides. A typical reduction and conditioning cycle involves filling the previously evacuated system with close to 1 atm of hydrogen, after which the traps, tubing, and other heat-resistant items are heated with a gas-air flame. The system is then evacuated and heated again to expel moisture. To passivate the apparatus, a mixture of 5–10% fluorine in an inert gas is first introduced, and nickel or Monel parts are heated. The gas is replaced by fluorine somewhat below 1 atm of pressure, and nickel and Monel parts are heated to a dull red until the pressure drop ceases. Sometimes this passivation is followed by another treatment with a highly reactive fluorine compound, such as ClF_3, which is introduced for a few hours to condition parts which cannot be heated, such as Teflon packed valves or fluorocarbon reaction vessels. The disadvantage of this last passivation step is that ClF_3 is strongly adsorbed on the walls of the apparatus, and extensive pumping is needed to remove it from the system. As a general rule, the conditioning should include a fluorinating agent which is at least as strong as any that subsequently will be handled. Some materials, such as stainless steel, platinum, and gold, do not form a protective fluoride film and therefore cannot be conditioned.

Two typical metal vacuum systems for fluorine compounds are illustrated in Fig. 10.15. In either system the vacuum pumps are protected from the vacuum

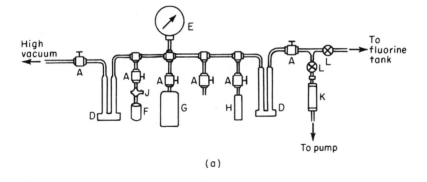

(a)

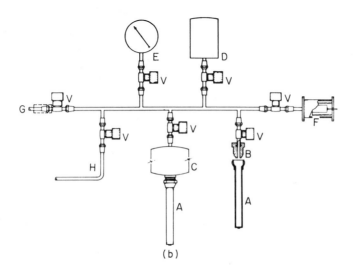

(b)

Fig. 10.15. Metal vacuum systems for handling fluorine and reactive fluorides. (a) A design used extensively at Argonne National Laboratory constructed of nickel tubing and Monel valves (A) with cone joints (illustrated in Fig. 10.13); (D) nickel U-trap; (E) Monel Bourden gauge (0–1000 torr); (F) 130-mL nickel reactor can (Fig. 10.17); (G) 1,500-mL nickel storage or measuring can, (H) 85-mL nickel can, (J) brass valve; (K) soda-lime trap to protect vacuum pumps; (L) Monel valve. (Reproduced by permission of the copyright holder, The University of Chicago Press, from *Nobel Gas Compounds*, H. H. Hyman (Ed.), Chicago, 1963.)

(b) A less massive vacuum system: (A) Kel-F or Teflon reaction tubes (illustrated in Fig. 10.16); (B) Flare fitting for the reaction tube; (C) nickel expansion can; (D) calibrated nickel can; (E) Monel Bourdon gauge; (G) to soda-lime traps and pumps; (H) to sampling cells. Monel bellows valves (V), Monel Swagelok joints, and nickel tubing are used for the main components in the vacuum line. (Reproduced from T. A. O'Donnel and D. F. Stewart *Inorg. Chem.* 5, 1434 (1966), by permission of the copyright holder, The American Chemical Society.)

229

line by means of a soda lime trap. Kel-F or Teflon reaction tubes are convenient because their translucence permits the observation of the contents. Details of the reaction tube are given in Fig. 10.16. The tube can be machined from Kel-F bar stock or it can be fashioned from heat-formable and -sealable Teflon FEP. The metal reaction can illustrated in Fig. 10.17 is preferred when the reactants do not have to be observed and when modest pressures of reactants are desired. For the manipulation of HF solutions, apparatus made from Kel-F or Teflon is desirable. A reaction vessel is illustrated in Fig. 10.18 and a cell for spectroscopy is shown in Fig. 10.19.

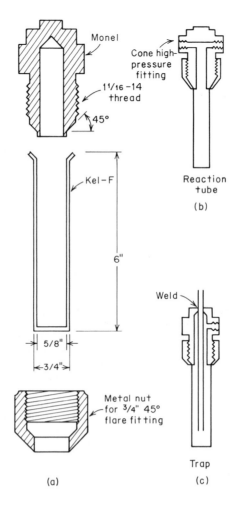

Fig. 10.16. Perfluorocarbon reaction tube and flange-fittings.) (*a*) Detailed cross-section of the components; (*b*) and (*c*) two adaptations of the basic design. The fluorocarbon tube may be machined from Kel-F; alternatively, Teflon FEP tubing may be heat molded to form a flange and heat crimp sealed at the base.

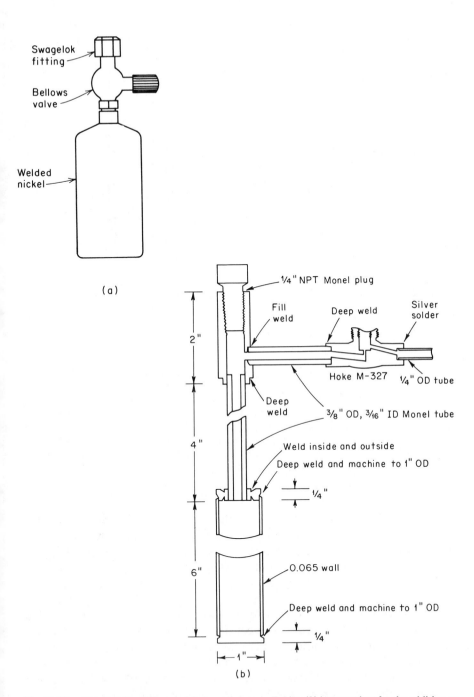

Fig. 10.17. Detail of two different nickel reaction cans. Design (*b*) is convenient for the addition or removal of solids. [Reproduced by permission of the copyright holder, The American Chemical Society. From W. E. Tolberg, R. T. Tewick, R. S. Stringham, and M. E. Hill, *Inorg. Chem.*, 6, 1156 (1967).]

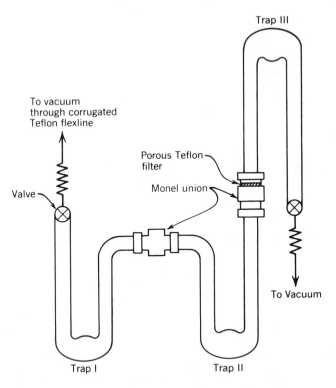

Fig. 10.18. Assembly for the manipulation of HF solutions. Monel Swagelok unions are used to join U-traps constructed from Teflon FEP tubing. A rigid Kel-F or Monel tube must be placed inside the Teflon tubing to make it sufficiently rigid to hold a swage-type seal. Similar Swagelok attachments are made to the valves. [Reproduced by permission of the copyright holder, The American Chemical Society. From K. O. Christe, C. J. Schack, and R. D. Wilson, *Inorg. Chem.*, 16, 849 (1977).]

Since fluorine and most volatile fluorides are corrosive and highly toxic, the vacuum rack for a fluorine handling system generally is placed in a hooded enclosure (Fig. 10.20) and the fluorine tank is often enclosed in a separate enclosure which can be manipulated remotely (Fig. 10.21).

B. Flow Reactors. Laboratory-scale catalytic reactors and reactors for the reaction of solids with gases are often constructed from metal. One of the principal objectives in the use of laboratory-scale catalytic reactors is the determination of rate data which can be associated with specific physical and chemical processes in a catalytic reaction. Descriptions are available for these kinetic analyses as they relate to reactor designs and reaction conditions.[9]

[9]E. G. Christoffel, *Catal. Rev.*, 24 159 (1982); L. K. Doraiswamy and D. G. Tajbl, *Catal. Rev.*, 10, 177 (1974); L. L. Hegedas and E. E. Peterson, *Catal. Rev.*, 9, 245 (1974).

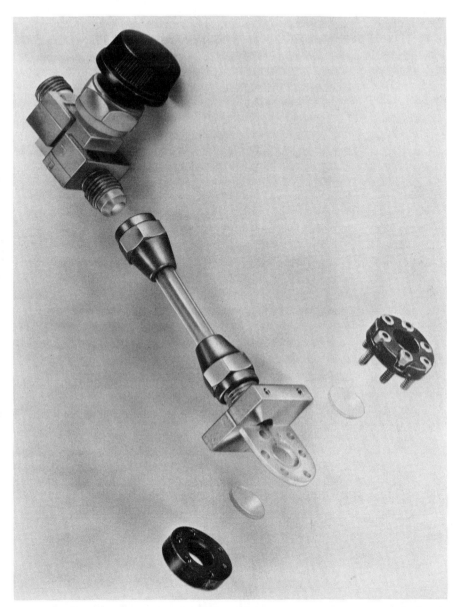

Fig. 10.19. Spectroscopic cell for HF solutions. All internal parts are Kel-F, Teflon, or, in the case of window material, sapphire or silver chloride. A commercial Teflon valve may be substituted for the custom Kel-F valve shown here. (Reproduced from H. H. Hyman and J. J. Katz *Nonaqueous Solvent Systems*, T. C. Waddington Ed., p. 47 (1965) by permission of Academic Press Inc.)

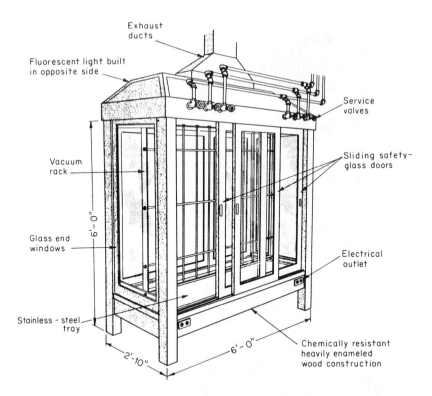

Exhaust
ducts

Fluorescent light built
in opposite side

Service
valves

Vacuum
rack

Sliding safety-
glass doors

6'-0"

Glass end
windows

Electrical
outlet

Stainless - steel
tray

Chemically resistant
heavily enameled
wood construction

2'-10"

6'-0"

Fig. 10.20. Hooded vacuum rack. (Reproduced from D. R. Ward, *Laboratory Planning for Chemistry and Chemical Engineering*, H. F. Lewis, (ed.) (1962) by permission of the copyright holder, Van Nostrand Reinhold Co., Inc., New York.)

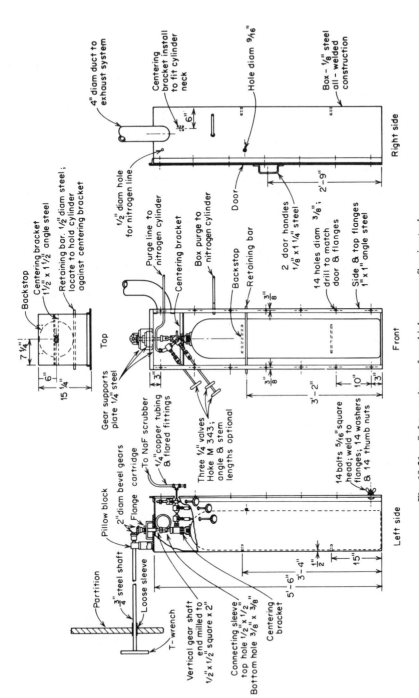

Fig. 10.21. Safety enclosure for a high-pressure fluorine tank.

235

GENERAL REFERENCES

Broker, W., and A. L. Mossman, *Matheson Gas Data Book*, Matheson Gas Products, P.O. Box 1587, Lyndhurst, NJ 07071. A general description of the handling of compressed gases is given along with information on fittings and properties of specific gases.

Calahan, F. J., *Swagelok Tube Fitting and Installation Manual*, Crawford Fitting Co., Cleveland. Procedures for forming and joining metal tubing are described.

Canterford, J. H., and T. A. O'Donnell, 1968, in H. B. Jonassen and A. Weissberger, Eds., *Techniques of Inorganic Chemistry*, Vol. 8, Wiley, New York, p. 291. General techniques for fluorine handling.

Hagenmuller, P., Ed., 1985, *Inorganic Solid Fluorides*, Academic, New York. Chapter 2 of this book, "Preparative Methods," gives an extensively referenced account of synthetic methods used in inorganic fluoride research. Several apparatus are pictured and described in detail.

Hyman, H. H., Ed., *Noble Gas Compounds*, 1963, University of Chicago Press, Chicago. Many chapters contain experimental detail.

Guide to Safe Handling of Compressed Gases, Matheson Gas Products, P.O. Box 1587, Lyndhurst, NJ 07071. The safe handling, transporting, and storage of compressed gases are described.

Moore, J. H., C. C. Davis, and M. A. Caplan, 1983, *Building Scientific Apparatus*, Addison-Wesley, Reading, Mass. Includes coverage of metalworking and the construction of apparatus from metal components.

Steere, N. V. Ed., 1971, *Handbook of Laboratory Safety*, The Chemical Rubber Co. Press, Cleveland. Contains a chapter on compressed gas cylinders and cylinder regulators.

Weinstock, B., *Record of Chemical Progress*, 23, 23 (1963). Apparatus for the synthesis and characterization of xenon fluorides is presented.

Safety

Many air-sensitive compounds are very strong reducing or oxidizing agents, and therefore care must be used in their manipulation to avoid explosions and fires. Careful planning and execution of experiments can minimize the danger of reactive compounds like diborane. However, there is an unfortunate tendency to forget mundane hazards such as the fire and explosion hazards of organic solvent vapors or hydrogen gas. The chemist also should be socially responsible by never discarding highly toxic materials down the drain and by chemically scrubbing noxious gases from gas streams *before* the gas is allowed to go into a hood.

I.1 COMBUSTIBLES

The *explosion limits* for vapor-air mixtures are the maximum and minimum concentrations of vapor between which a flame will propagate through the mixture. Large quantities of flammable solvents and gases must be handled in well-ventilated areas so that the vapors are dilute. Laboratory refrigerators present an explosion hazard because the storage of highly volatile solvents and volatile solutions in them may lead to the buildup of the solvent vapor concentration. The usual household refrigerator is equipped with a thermostat and a light switch which may spark and ignite solvent vapors. Explosion-proof refrigerators are commercially available.

High pressures of oxygen may spontaneously ignite many combustible materials such as hydrocarbon greases and oils. Therefore, regulators, gauges, and other oxygen-handling systems should be strictly oil-free.

I.2 UNSTABLE COMPOUNDS

A large number of compounds are unstable with respect to spontaneous decomposition. Only a few of the most important will be listed here.

1. Acetylene may spontaneously explode if its pressure exceeds 15 psig. Commercial acetylene cylinders are filled with a porous material soaked in acetone which maintains a safe acetylene pressure by dissolving the gas. Acetylene should never be passed through a vacuum pump, which will compress the gas.
2. Acetylides of heavy metals are often explosive.
3. Azides, diazo compounds, and fulminates are frequently explosive.
4. Nitrogen chlorides, bromides, and iodides are endothermic compounds which may detonate.
5. Organic peroxides are liable to detonate. Exposure of ethers, alkenes, and alkynes to air and light leads to peroxide formation. Distillation of these solvents poses an increased danger due to the concentration of peroxides in the still pot as the distillation proceeds. When present in small amounts, the peroxide is removed before distillation by the use of reducing agents such as ferrous sulfate in acidic aqueous solution. Peroxides may be detected by the addition of a few drops of aqueous 10% KI solution to a few mL of the ether. The presence of a peroxide is revealed by the appearance of the brown triiodide color.
6. Liquid or solid ozone is explosive.
7. Perchlorate salts of metal complexes are liable to detonate. Perchlorate has been a popular counter ion for metal complexes, but safer ions, such as tetrafluoroborate or hexafluorophosphate, can generally take its place.

I.3 SOME DANGEROUS MIXTURES

Undiluted mixtures of strong oxidizing agents with strong reducing agents are potentially dangerous, as are mixtures which lead to unstable compounds. A few of these are listed below.

1. Perchloric acid forms explosive mixtures with organic compounds and easily oxidized inorganic compounds such as ammonia.
2. Nitric acid with organic compounds and ammonium nitrate with metal powders or organic materials are potential explosives.
3. Ammoniacal silver(I) solutions may form an explosive component after long standing or in the presence of strong base.
4. In the presence of air, mercury and anhydrous liquid ammonia form an explosive compound, as does yellow mercuric oxide with ammonia.

5. Activated charcoal with nitrates, chlorates, liquid air, or liquid oxygen forms explosive mixtures.

6. Liquid air or liquid oxygen with organic matter is very dangerous.

7. Active metals, such as the alkali metals, alkaline earths, and finely divided aluminum, may explode when mixed with chlorinated hydrocarbons, such as chloroform and carbon tetrachloride.

8. Potassium permanganate with sulfuric acid yields the explosive compound permanganic acid.

9. Decaborane (and possibly some other boron hydrides) in the presence of carbon tetrachloride, and possibly other halogenated hydrocarbons, forms a very shock-sensitive high explosive.

10. Similarly, an explosive material results when alkyl phosphines are mixed with halogenated hydrocarbons.

11. Sodium amide ($NaNH_2$), forms an explosive compound upon prolonged exposure to the air.

12. The distillation of ethers from lithium aluminum hydride occasionally leads to an explosion. The exact cause is not known, but CO_2 may be involved. The danger can be minimized by predrying the ether with calcium hydride and then using a minimum amount of $LiAlH_4$ for final distillation. Also, the distillation should be performed behind a blast shield, and the still pot should never be allowed to go dry. Frequently, a safe but powerful desiccant, such as benzophenone ketyl or sodium-potassium alloy, may be used in place of $LiAlH_4$.

13. The reaction of potassium with atmospheric oxygen leads to a surface layer of potassium superoxide. When this oxidizing layer is embedded in the metal, an explosion may result. Therefore, the oxidized surface layer must be scraped off before the metal is cut. Presumably a similar problem exists with rubidium and cesium.

14. Dimethyl sulfoxide is reported to form an explosive mixture with $B_9H_9^{-2}$ and with diborane. It is probable that other boron hydrides and hydroborates behave similarly.

15. N_2O_4 with some chlorinated hydrocarbons and olefins forms shock sensitive compounds.

I.4 DISPOSAL OF REACTIVE WASTES

Lithium aluminum hydride wastes may be destroyed by passing nitrogen through a water bubbler, an empty trap to catch the water, and then into the flask containing the waste. Alternatively, hydride wastes may be covered with rags and burned in an open area. Aluminum alkyls may be burned in the open, or a dilute solution of the alkyl in a hydrocarbon may be run into agitated water in a vessel which is blanketed with nitrogen. Variants of these methods are effective with a wide variety of compounds which are decomposed by water or alcohols. Alkali metal residues may be destroyed by isopropyl or t-butyl alcohol and then flushed down the drain with water. Since potassium is more difficult to destroy completely by alcohols than is sodium, extra care must be used when disposing of potassium.

Nitrogen halides are destroyed with cold base. Azides and fulminates may often be destroyed with acid, while heavy metal acetylides are decomposed by ammonium sulfide. The removal of peroxides by reduction has been described above.

1.5 HIGH-PRESSURE GAS CYLINDERS

Procedures for safe handling of pressurized gases are given in Chapter 10.

1.6 ASPHYXIATION BY INERT GASES

An atmosphere of inert gases will not support life; therefore, good ventilation is required.

1.7 EXTINGUISHING FIRES

Carbon dioxide fire extinguishers are widely used in laboratory fire fighting because the CO_2 does not damage apparatus or cause electrical shorts. Furthermore, carbon dioxide is effective against a wide variety of fires and is not toxic. The National Fire Protection Association adopts the following classification of fires:

Class A fires occur with wood, paper, fabric, and similar solids. Water and wet fire extinguishers are recommended because they are effective in quenching burning embers.

Class B fires occur with liquid hydrocarbons and similar combustible liquids. Foam, dry-chemical, halocarbon, and CO_2 extinguishers are recommended.

Class C fires occur in live electrical equipment. Clearly, water and wet-type fire extinguishers must be avoided.

Class D fires involve strong reducing agents such as active metals (magnesium, titanium, zirconium, and alkali metals), metal hydrides, and organometallics. Special dry-chemical fire extinguishers are available for these fires (e.g., Ansul Co.). Sand is also useful for small fires of this type. Water should be avoided because it promotes the fire by liberation of hydrogen or hydrocarbons.

GENERAL REFERENCES

Bretherick, L., 1985, *Handbook of Reactive Chemical Hazards*, 3rd ed., Butterworths, Boston. Chemical hazards, with references for 4,900 compounds.

Keith, L. H., and D. B. Walters, Eds., 1985, *Compendium of Safety Data Sheets for Research and Industrial Chemicals*, 3 Vol, VCH Publishers, Deerfield Beach, Florida. Health hazards, first aid, recommended storage, and references for 867 chemicals.

Prudent Practices for Handling Hazardous Chemicals in Laboratories, National Academy Press, Washington, D.C., 1981. General laboratory practice for hazardous materials, electrical hazards, laboratory ventilation, storage of chemicals, disposal of chemicals, and threshold limit values for chemical substances.

Sai, N. I., Ed., 1984, *Dangerous Properties of Industrial Materials*, 6th ed., Reinhold, New York. Physical properties, fire and explosion hazards, toxicity, and incompatibility for about 20,000 chemicals.

Steere, N. V., Ed., 1971, *Handbook of Laboratory Safety*, The Chemical Rubber Co., Cleveland, Ohio.

A P P E N D I X II

GLASS AND GLASSBLOWING

Even though the services of a professional glassblower may be available, it is often much more efficient for the chemist to construct, modify, and repair small glass items, and most chemists involved in vacuum line work are capable of much more. In addition to the advantage of efficiency, glassblowing affords an agreeable diversion which is doubly rewarding when an aesthetically pleasing apparatus is produced. As an aid to the design of glass apparatus, this appendix presents information on the properties of some common laboratory glasses, followed by instructions for some basic glassblowing operations. These instructions should be supplemented, if possible, by observing an experienced glassblower. The beginner will find practice necessary to develop the manual dexterity and judgment which are required in glassblowing.

II.1 PROPERTIES OF GLASSES

The glasses in general laboratory use contain SiO_2 and/or B_2O_3 as the glass-forming constituent and frequently a metal oxide modifier, which is added to lower the viscosity of the glass or to impart other desirable characteristics. Glass is an amorphous substance which may be viewed as a highly viscous liquid. Thus, unlike a pure crystalline substance, it does not display a sharp melting point, but despite this, it is convenient to speak of temperatures around which recognizable changes in properties occur: The *strain point* corresponds to the temperature at which internal stress is largely relieved in 4 h; more precisely, it

242

corresponds to a viscosity of $10^{14.5}$ poise (P).[1] At the *annealing point* ($10^{13.0}$ P), internal stress is substantially relieved in 15 min. The *deformation point* corresponds to a temperature at which viscous flow neutralizes the effect of thermal expansion (about 10^{11}–10^{12} P). The *softening point* has an involved definition, but roughly speaking it corresponds to a temperature at which a small heated rod will slowly elongate under its own weight. This corresponds to a viscosity of approximately $10^{7.6}$P.

Glass is generally annealed at a temperature between the strain point and the annealing point and is worked above the softening point. The relationship of these temperatures is brought out more clearly in Fig. II.1, where it may be seen that fused quartz is the most refractory of the common laboratory glasses, and Vycor, a 96% silica glass manufactured by Corning Glass Works, is somewhat less so. Because of the high softening points of these two glasses, they must be worked with a hydrogen-oxygen flame and, unlike borosilicate glass, they are generally blown while in the flame. The ability of these glasses to withstand high temperatures leads to their use in combustion tubes and similar applications. The maximum temperature for continuous use is about 900°C for Vycor, while

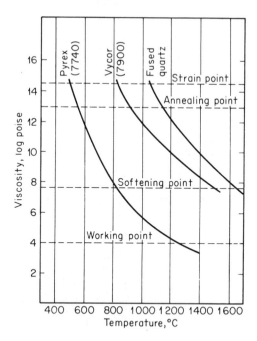

Fig. II.1. Viscosity–temperature curves for some common laboratory glasses. The numbers in parentheses correspond to Corning designations. (Adapted from Corning Glass Works, Corning, NY, Bulletin B-83, 1957.)

[1]These are tentative designations specified by the American Society For Testing Materials.

continuous heating of fused quartz above about 1000°C will lead to failure due to devitrification (crystallization). Also, because of their high strength and very low coefficient of expansion, these glasses can withstand considerable thermal shock without breaking. For example, a red-hot tube can be plunged into water without cracking. Another important property is their transmission of ultraviolet and near-infrared radiation (Fig. II.2); accordingly, fused quartz and, to a lesser extent, Vycor are often employed in cell windows and reaction vessels for photochemical work. Finally, the high strength of fused quartz leads to its frequent use in fibers for the suspension of galvanometer parts, as the helical spring for weighing small samples, and for spiral gauges.

The majority of chemical glassware is constructed from borosilicate glass, which has the advantages for routine use of a reasonably low coefficient of expansion and a convenient softening temperature for glassblowing. In addition to the oxides of silicon and boron, these glasses contain small amounts of the oxides of aluminum, sodium, and potassium. The trade designations for the glass used in most laboratory equipment are Pyrex 7740 (Corning Glass Works) and Kimax K-33 (Owens-Illinois). They are very similar in composition and are compatible with each other, but cannot be fused directly to flint glass or fused quartz. Although the softening temperature is 820°C, it is unwise to use borosilicate glass

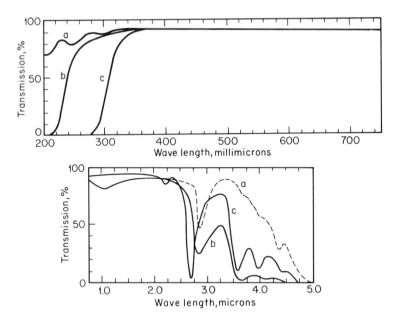

Fig. II.2. Transmission of Pyrex, Vycor, and fused quartz. Transmission curves for 2-mm-thick samples of (a) fused quartz (GE type 102), (b) Vycor (Corning 7910—note that this is a specially controlled grade; the more common 7900 has an ultraviolet cutoff similar to 7740), and (c) Pyrex (Corning 7740). (Adapted from data supplied by Corning Glass Works and General Electric Co.) Certain glasses and types of fused quartz are available which lack the infrared absorption at about 2.7 microns.

above 500°C, especially if evacuated or under stress.[2] Unless otherwise stated, the pieces of apparatus described in this book, as well as the glassblowing operations which follow, apply to Pyrex 7740 or Kimax K-33. Because of its poor resistance to thermal shock and low softening temperature, flint glass is not desirable for most laboratory apparatus. The junction of dissimilar glasses is generally performed through a graded seal composed of rings of compatible glasses which have a smooth variation in expansion coefficient and melting point. The very useful Pyrex-to-quartz graded seals are available commercially. Also, a serviceable union may be made between Pyrex and the high-temperature-resistant ceramics.[3]

The tensile strength of glass is influenced by a sufficient number of variables, so it is difficult to specify values which will apply to laboratory conditions. It is decreased by surface scratches, moisture, and many solvents. Also, the strength decreases for glass under continual load. As a result, strengths greater than 1×10^6 lb/in.[2] and as low as 19 lb/in.[2] have been determined. It has been suggested that a value of 1,000 lb/in.[2] is reasonable for annealed glass.[4] Accordingly, this value can be used to calculate the values for the bursting pressures of glass tubes. Any glass tube under pressure must be treated as dangerous because of the highly variable tensile strength of glass. Values calculated for sealed tubes are of limited value because tests have shown that the weak point is usually the glass seal.[5]

II.2 EQUIPMENT AND MATERIALS

An amazing variety of glass apparatus can be constructed by using simply a torch and low-pressure natural gas and oxygen supplies. Also, a pair of didymium-tinted glasses is a necessity to eliminate the yellow glare of sodium emission which occurs when borosilicate glass is heated.

A more representative group of equipment for use in the laboratory includes the following items, some of which are illustrated in Fig. II.3:

Hand torch[6] and oxygen pressure regulator

Didymium glasses

Glass knife or three-cornered files

---

[2]A. W. Laubengayer, *Ind. Eng. Chem.*, 21, 174 (1929), describes some experiments which give an indication of the useful temperature range for Pyrex.

[3]J. A. Alexander, *Fusion*, 3, 5 (1956); E. R. Nagle, *Fusion*, 10, 23 (1963).

[4]*Properties of Selected Commercial Glasses*, Bulletin B-83, Corning Glass Works, Corning, N.Y.

[5]J. Tanaka, *J. Chem. Educ.*, 50, 4978 (1972).

[6]Surface-mix hand torches are available from Bethlehem Apparatus, Inc., Hallertown, Pa. Internal-mix hand torches are available from most laboratory or welder's supply houses. A range of tips is also required with the internal-mix torches.

Swivel and blowhose

Tweezers

Tapered graphite rods (hexagonal or circular cross section)

Set of plain and one-hole stoppers and corks

Glass fiber paper and/or aluminum foil

Polariscope, or two pieces of Polaroid film and a flashlight

Caliper scale

A pair of rollers mounted on a stand

A small fireproof bench top or table

A reasonably dust-free cabinet stocked with a variety of standard wall tubing, heavy-walled capillary tubing, and rod

A ring stand and clamps with woven-glass-covered fingers

Surface-mix bench burner for large items

Clean glass is used for glassblowing because foreign matter is easily incorporated into molten glass. Aside from the obvious bad effects this can have on the surface and character of the glass, specks of dirt are a frequent source of pinholes. To avoid picking up foreign matter, it is desirable to use freshly prepared surfaces for making joints and to minimize contact of these surfaces with anything but glass or clean glass-working tools. In this connection, tubing should be stored in a cabinet to avoid the accumulation of dust.

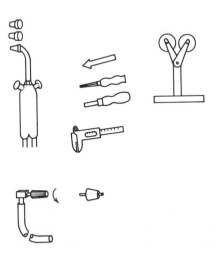

Fig. II.3. Some glassblowing implements. An internal-mix hand torch and a series of tips are shown in the upper left. In the upper right is a set of rollers. From top to bottom in the center are a brass-shaping tool, a tapered carbon rod on a handle, a glass knife with tungsten-carbide cutting edges, and a vernier caliper scale. At the bottom a swivel and blowhose are illustrated at a slightly larger scale than the other items. The short length of tubing on this swivel may be connected directly to an item or it may be attached by means of a one-hole stopper with glass tube. In either case free rotation of the item and simultaneous blowing are possible.

II.3 SEQUENCE OF OPERATIONS

Molten glass resembles very thick molasses, so it tends to sag and flow under the influence of gravity. Therefore, whenever possible, the pieces being worked are rotated to avoid the pileup of glass. When clamping is necessary, the effects of sagging may be minimized by clamping tubes in a vertical fashion and placing a tube with thick walls above one that has thin walls.

In general, it is best to finish a piece without interruption and to anneal it immediately with the hand torch. An even wall thickness is always desirable because thin glass is weak, while lumps are frequently strained. Before an intricate apparatus is constructed, the sequence of work is planned so that it is always possible to apply pressure (blow) on a joint that is being constructed. Occasionally it is necessary to blow a temporary hole in an apparatus to serve as a place to attach a blowhose. For example, suppose a bulb with an attached stopcock is being constructed from a round-bottom flask and a stopcock. A small hole is first blown into the bottom of this flask (Fig. II.4a), and either a blowhose with a swivel is attached or else a temporary length of tubing is attached. Either of these provides a mechanism for sealing off the neck and then blowing another hole which will match the stopcock (Fig. II.4b). After the stopcock is attached, it provides an opening for blowing while the temporary hole is being sealed (Fig. II.4c). Details of glassblowing operations involved in the construction of this bulb are presented in the following sections.

II.4 ANNEALING

In the process of working glass, the thermal gradients introduce strains which must be relieved if reliable service is expected from the glassware. The most sat-

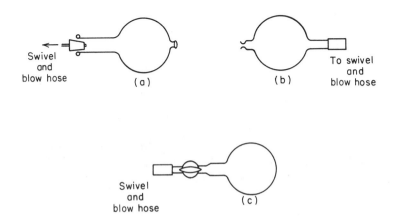

Fig. II.4. Some steps in attaching a stopcock to a flask. This sequence allows the glass to be blown at each stage.

isfactory procedure involves the use of a glassblower's annealing oven, which provides gradual and well-controlled warming and cooling cycles. However, large and intricate parts must be annealed with the bench torch immediately after construction and then placed in an annealing oven. A muffle furnace will serve well as an annealing oven. For most laboratory items of moderate thickness, the furnace and contents are allowed to warm from room temperature to 570°C; this temperature is maintained for half an hour, and then the furnace is turned off and allowed to cool slowly. Many small parts may be adequately annealed with the bench torch alone. A cool, bushy flame is used, and attainment of the annealing temperature is judged by the appearance of the yellow sodium flame. Strains tend to concentrate on either side of a section which has been worked, so a generous area should be annealed. After the glass has been thoroughly and evenly heated to the annealing temperature, the flame temperature is gradually reduced. Some glassblowers recommend that the last bit of heating should be done with a sooty flame; however, it is doubtful that such a cool flame is worthwhile. For large parts, it is handy to have a Fisher burner available to use in the annealing process. After the glass has cooled, strains may be located as light streaks when the apparatus is viewed between crossed Polaroids.

When a large apparatus is mounted on a lattice, strains are inevitably introduced by the clamping operation. This is partly avoided by using a minimum number of clamps. Also, all heavy items, such as those which contain appreciable quantities of mercury, must be individually supported. The clamp fingers are best covered with woven glass sleeves or a layer of glass fiber paper. These covering materials have a degree of resiliency and allow the use of a torch in the vicinity of the clamp for the purpose of making repairs or alterations, and for flaming out and annealing the apparatus. After a vacuum line has been assembled, but before it is evacuated, the glass is softened and annealed at selected points to remove strains introduced by the clamping.

II.5 CUTTING GLASS TUBING

Glass is generally broken and not cut. Small-diameter tubing is conveniently broken by making a scratch about one-fourth of the way around the tube and then grasping the tube on either side of the scratch and simultaneously pulling and bending. The nature of the scratch is quite important; it should be made with a sharp tool and with a smooth, firm motion, so that the glass crackles. Sawing at the glass with the corner of a dull file produces a rather ineffective scratch. The breaking operation is facilitated if the scratch is moistened. With large tubing or with a permanently clamped piece, this method cannot be used, and in these cases the finite thermal expansion coefficient of Pyrex is utilized to produce a strain around the scratch. The simplest procedure is to heat the end of a small-diameter glass rod until it is white hot and then quickly touch this to the glass just beyond the scratch. If the resulting break is not sufficiently long, it can be led around the tube by touching the hot rod just beyond the end of the break.

A variant of this technique is to extend the scratch about three-fourths of the way around the glass and then to direct a small but very hot flame at right angles to the tube so that it grazes the tubing along the middle of the scratch. While a smooth, clean break can be produced by the foregoing methods, a more satisfactory surface for subsequent glassblowing is often produced by a fire cutoff, which is described in Section II.7. It is sometimes handy to smooth up a jagged end or remove glass by brushing the end with a metal screen. However, this practice is not recommended if the end is going to be glassblown, because metal particles adhere to the glass.

II.6 BENDING GLASS TUBING

Simple bends can be made by evenly heating a broad section of tubing and allowing the free end to sag under its own weight. For this purpose a large flame is used which is sufficiently hot to surpass the softening point, but not hot enough to melt the glass. A bend may also be made by evenly heating the desired length of tubing to the softening point and then removing it from the flame and making the bend. Kinks or very sharp bends should be avoided because they break easily under strain. It is sometimes not possible to avoid some collapse of a tube when a bend is made, but its shape is restored by stoppering one end of the tubing, heating the section to be re-formed to the working point in a large flame, and then gently blowing on the open end of the tubing just after the flame is removed. With a little practice, one can combine the bending and blowing into one continuous operation.

II.7 TEST-TUBE ENDS AND FIRE CUTOFFS

Both ends of a tube are grasped and rotated in a hot flame until the glass becomes molten (Fig. II.5a). The tube is then pulled apart in the flame, and excess glass is removed from one half by melting the unwanted glass and pulling it out with a piece of cool rod or tubing (Fig. II.5b). This should leave a completely closed tube, free of excess glass. The closed end is now heated in a hot flame until it becomes molten, and during this heating, the sagging of glass is prevented by constantly rotating the tube (Fig. II.5c). As the tube is being rotated, it is removed from the flame and gently blown out to give a hemispherical end of uniform thickness (Fig. II.5d). Timing is important in this process because if the end is blown out immediately after heating, any thin regions will blow out first; but with the correct delay, thick sections which cool the slowest will be preferentially blown out to yield a uniform wall thickness. Failure to rotate the glass constantly is a common error, which leads to uneven heating of the end and a lopsided product. Another error is to heat the closed end too long, which leads to an excessive accumulation of glass.

When one is forced to work close to the end of a tube, it is not possible to

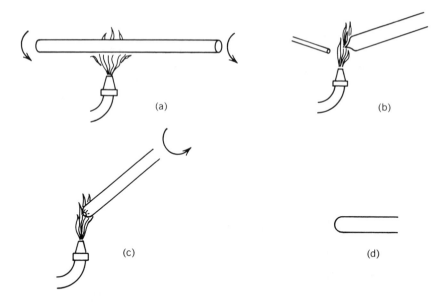

Fig. II.5.   Construction of a test-tube end. (*a*) A small section of the tube is melted and the halves pulled apart. (*b*) A small hot flame is used to melt excess glass, which is then pulled off with a cold piece of glass. (*c*) The end is heated evenly, removed from the flame when it is molten but before excess glass accumulates, and gently blown. (*d*) A finished product. The end should be hemispherical and the wall thickness uniform.

grasp both ends as decribed above. In this situation another piece of tubing or a rod is sealed to the end of the tube to serve as a handle. A very crude junction will suffice since the handle is temporary.

The preliminaries to produce a fire cutoff are the same as for a test tube end. Instead of gently blowing the molten glass into a hemisphere, a more vigorous puff is given to produce a large, thin bulb. Just as with the test tube end, even heating is necessary to ensure a well-shaped opening (Fig. II.6). If properly blown, the thin glass is easily brushed off by use of glass rod or another piece of glass tubing to leave an even opening with a slight flare and a thin edge.

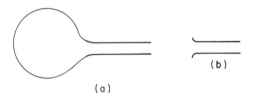

Fig. II.6.   A fire cutoff. (*a*) A large, very thin bulb is blown out. (*b*) The thin glass is brushed away.

II.8 END-TO-END SEALS

The end-to-end seal is a very common operation in constructing glass equipment. The process is aided if the ends to be joined are smoothly broken or are
evenly blown out; and it is usually desirable to start with freshly broken or fire-
cut pieces because this minimizes the chance of incorporating dirt into the joint.

When the joint is to be made offhand, the torch is clamped and a free end of
one of the tubes is stoppered; the tubes are then grasped in either hand and
rotated in a hot and rather large flame (Fig. II.7a). This process requires some
dexterity, since the rotation must be even and the ends must not touch but must
become thoroughly molten. The heating is carried on at an even but quick pace
to avoid a large accumulation of molten glass around the two ends. Next, the
pieces are removed from the flame and pressed together. This is generally followed by a gentle pull and simultaneous blowing to thin out the ring of glass
which has formed at the junction (Fig. II.7b). Here again, it is necessary to develop some dexterity so that the pieces are initially joined in a concentric fashion.
If the initial union is badly skewed or if a continuous hole-free seal is not
achieved, a smoother final product is generally obtained and time is saved by
starting over, rather than attempting to patch up the joint.

The pulling-out and blowing operations described above are not absolutely
essential, because the next step is to smooth out the joint. This may be accomplished in several ways. Generally, the part is rotated in a hot flame until the
glass in the joint is thoroughly melted; it is then removed from the flame and
gently pulled to thin out the thick glass and blown to restore the glass to its
correct diameter. Close coordination of the rotation rate for each hand is necessary to avoid twisting the glass during the heating. When unwieldy parts are

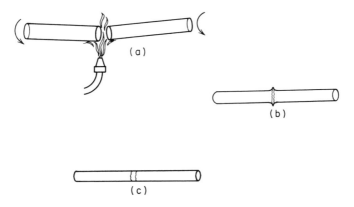

Fig. II.7.   Construction of an end-to-end seal. (a) A large hot flame is used to create an even molten
rim on both pieces of glass. (b) The two ends are joined, which leaves a small ridge of glass that must
be worked in. (c) The finished product should have a reasonably uniform wall thickness and no
vestiges of the ridge.

encountered, a sightly different approach may be taken. In this case, the welt of glass at the junction is worked in by heating a small segment of the glass to a molten state and then blowing it out to the correct radius. One then moves on to the adjacent segment and works it in. A final heating of the joint to the softening point, alignment, and blowing may be necessary; but the above process does avoid a continuous belt of molten glass, which may be hard to handle because of the bulky nature of the apparatus.

An end-to-end seal may also be made by firmly clamping one of the parts, preferably in a vertical direction, and lightly clamping the other so that the edges are aligned but do not quite touch. One hand is used to steady the lightly clamped piece and the other directs the hand torch around the two ends to give continuous molten edges. The lightly clamped piece is then pressed into the other, and the parts are worked together as described above. If the glass sags too much toward the lower component, the part may be inverted and additional smoothing performed.

When the two tubes to be joined are of different diameters, the task is slightly more difficult. In this case, a test tube end is formed on the larger tube; then a very small flame is directed in the center of this end, and only a small hole with a diameter which matches the small tube is blown out. To allow better control of the hole size and to yield a lip which facilitates the subsequent sealing, it is helpful to perform this blowing-out operation in two stages (Fig. II.8).

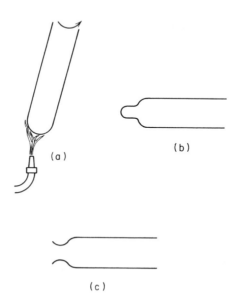

Fig. II.8.   Preparation of a large-diameter tube for an end-to-end seal with a smaller tube. (a) A test tube end is formed on the large tube, and a small molten spot is created on the end with a small hot flame. (b) A bulge is blown in the end. (c) The end of this bulge is melted and blown completely out. The diameter of this opening should match that of the small tube.

When an end-to-end seal is made with heavy-walled capillary tubing, the end(s) of the capillary tubing is blown out to produce an opening with a normal wall thickness. It is convenient to accomplish this in two stages, as illustrated in Fig. II.9. Subsequently, the seal is made in an ordinary manner.

II.9 T-SEALS

A round hole is blown in the side of a tube by first blowing a bulge and then concentrating the flame on the end of this bulge and blowing it out. This two-step operation gives a round hole and leaves a lip which is convenient for the subsequent sealing operation (Fig. II.10). The two edges to be sealed are heated evenly to the molten state, joined, and then worked in like the end-to-end seal. When a T-seal is made by joining a small tube to the side of a large one, care must be taken to avoid sudden cooling and reheating of the joint before it is finished and completely annealed. If this is not done, considerable strain develops because of the unsymmetrical configuration around the axis of the large tube.

II.10 RING SEALS

There are two common ways of making ring seals. One of these is illustrated in Fig. II.11, where it may be seen that a test tube end is formed on the outer tube,

(a)   (b)

Fig. II.9. Preparation of capillary tubing for an end-to-end seal. (*a*) A tube blown out to normal wall thickness. (*b*) The same tube with the end blown out.

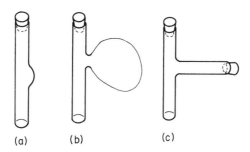

(a)   (b)   (c)

Fig. II.10. Construction of a T-seal. (*a*) A bulge is blown in the side of the tube. (*b*) The end of the bulge is heated and blown out to leave a small lip. (*c*) A junction is made in a manner similar to the construction of an end-to-end seal.

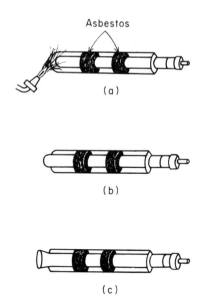

Fig. II.11.   Some steps in making a ring seal. (*a*) The formation of a seal between the inner and outer tubes. (*b*) A bulge is blown, which creates an extension of the inner tube. (*c*) The end of the bulge is blown out. This piece is now sealed to a tube of the same diameter as the inner tube.

and a smaller tube is then slipped inside and held centrally by glass paper. This inner tube is butted firmly against the test tube end, and a seal is made between the two tubes by playing a small and rather hot flame on the outside of the test tube end in the region where the two tubes touch. When the seal is made, a very small flame is directed in the center of the seal, and the glass is blown out by applying pressure through the inner tube. This is best accomplished by first blowing a bulge and then heating the end of the bulge and blowing it out. The resulting lip is then immediately sealed to a tube with a diameter similar to the inner tube, and the joint is annealed.

The second method (Fig. II.12) involves blowing a hole in the test tube end of a large tube. This hole should be just large enough to allow the central tube to slip through it; and it may be necessary to use a tapered carbon rod on the hot glass to obtain the correct size opening. A small annular bulge is then blown in the small-diameter tube in the region in which the seal is to be made, and this tube is slipped though the hole so that the bulge is on the outside. The concentricity of the tubes is ensured by several turns of glass paper around the central tube or by an appropriately bored cork. A small, hot flame is directed at the bulge where it contacts the hole, and the parts are rotated until the glass is molten. The small tube is then pushed toward the larger one to make the seal and then pulled back and blown to create a uniform wall thickness.

Ring seals are particularly susceptible to breakage from thermal strains, so it

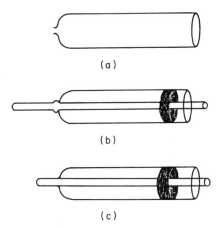

Fig. II.12. Second method for the construction of a ring seal. (*a*) Test tube end on a large-diameter tube has a small-diameter fire cutoff. (*b*) A small tube with a bulge is inserted. (*c*) The two tubes are sealed.

is best to make the complete seal without interruption and to anneal the final product carefully before it is allowed to cool.

II.11 CLOSED CIRCUITS

It is sometimes necessary to make a seal between two stationary tubes. Each particular case will require a slightly different solution, but one approach is to soften and stretch one piece of glass so that it meets the other, and then make the seal by heating the glass at the joint, pulling it together over any small gaps with a small piece of glass rod. It is also possible to join two tubes which do not meet by filling the space between them. When this is done, the flame is directed primarily onto the glass rod which is being used as filler and not on the open edges, because these tend to sag away from the joint if they are strongly heated. The same situation is encountered when a hole is being filled.

II.12 METAL-TO-GLASS SEALS[7]

A small tungsten electrical lead (1/8-in. diameter or less) can be sealed directly to Pyrex by the following procedure. A short length of Pyrex tubing is found

[7]Detailed information on glass-to-metal seals may be found in the following references: J. H. Partridge, *Glass-to-Metal Seals*, (Sheffield, U. K.: The Society of Glass Technology, 1949); W. H. Kohl, *Materials and Techniques for Electron Tubes*, (New York: Reinhold, 1960); Chapter 13; F. Rosebury, *Handbook of Electron Tube and Vacuum Techniques* (Reading, Mass: Addison-Wesley, 1965), pp. 54ff.

which will just slip over the wire and which is sufficiently short to leave adequate bare wire on either end for electrical connections. The tungsten wire is heated white hot, cooled, sanded until it is bright, and then carefully oxidized by quickly passing it through a flame. The oxide coating should be thin enough to give the appearance of gun-metal blue. The small piece of tubing is slipped over this tungsten wire and heated to collapse it onto the wire. If the seal is properly made, the tungsten-glass interface will have a copperlike appearance and few bubbles. The glass sheath may then be sealed into a hole in the apparatus. Generally, the hole is blown with a small lip which may be softened and then collapsed onto the glass sheath by tweezers or a carbon rod.

One of the most satisfactory metal-to-glass seals involves an Fe-Ni-Co alloy called Kovar. The coefficient of expansion of this alloy is close to that of Pyrex, to which it can be directly bonded; however, an even better match in expansion coefficients is obtained with a special borosilicate glass (Corning 7052, Kimble EN-1 or K-650). A variety of tubes is available commercially involving a Kovar tube sealed to a Pyrex tube through one or more intermediate glasses. The properties of Kovar are discussed in Appendix IV under iron (Section IV.1.E).

Copper-to-glass seals (Housekeeper seals) and stainless-steel-to-glass seals can be made and the directions may be found elsewhere.[8,9] These seals are somewhat strained and are generally not suitable if the joint is exposed to wide temperature variations.

It is possible to solder most metals directly to glass by use of indium metal (mp 155°C) or an indium alloy such as In 95 % Ag 5 % (mp about 145°C). The heated glass is rubbed with the molten solder to wet the surface and then heated in contact with the metal, which was previously "tinned" with the solder.[10]

A number of low-melting glasses have been specially formulated for "soldering" glass to glass, metal, or ceramics.[11] In general, the coefficient of expansion for the solder glass should nearly match that of the items being bonded.

II.13 HEALING CRACKS AND PINHOLES

Occasionally a crack will develop in an apparatus because of mechanical shock or internal strains which were not completely removed by annealing. Providing the glass is not shattered or the part is not under a large, externally produced strain, the crack may be healed by gentle heating. The area around the crack is first slowly heated, and then the soft flame (somewhat hotter than an annealing

[8]W. G. Housekeeper, *J. Am. Inst. Elec. Eng.*, 42, 954 (1923); also E. L. Wheeler, *Scientific Glassblowing*, (New York: Interscience, 1958), pp. 162ff.

[9]S. O. Colgate and E. C. Whitehead, *Rev. Sci. Instr.*, 33, 1122 (1962); J. E. Benbenek and R. E. Honig, *Rev. Sci. Instr.*, 31, 460 (1960).

[10]R. B. Belser, *Rev. Sci. Instr.*, 25, 180 (1954).

[11]Solder Glass Kit, Owens-Illinois Co., Toledo, OH; Quartz Cement Binder, American Thermal Fused Quartz Co., Montville, NJ.

flame) is moved gradually toward the crack. In many cases this process will lead to the disappearance of the crack. If, however, the crack does not heal, it will slowly open up as the glass sags away from the crack. A piece of cane (small-diameter glass rod) may then be used to unite the open edges of the crack and some final blowing will be necessary to obtain a smooth seal. A pinhole usually is sealed by melting the spot with a small hot flame and pulling away a small amount of glass with a glass rod.

II.14 SEALING TUBING UNDER VACUUM

Operations such as the collection of a sample in a sealed vial or the performance of reactions in sealed tubes require the ability to seal off tubing under vacuum. This is easily accomplished if a thickened section of glass with a moderate inside diameter (about 5 mm) was included in the apparatus (Fig. II.13*a*). The thickened section should be sufficiently long so that the adjacent thinner tubing is not softened during the seal-off operation. The inner wall of the constricted portion should be free from foreign matter, and throughout the sealing operation the

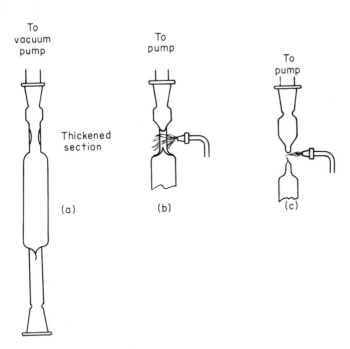

Fig. II.13.   Vacuum seal-off. (*a*) The tube is evacuated. If a volatile substance is being sealed into the tube, the lower end of the tube is immersed in liquid nitrogen. (*b*) Thickened section collapsed by a soft flame. (*c*) A hot flame is played on the collapsed section, and the sealed tube is pulled away. (*d*) Not shown. The tube end is gently annealed.

vessel is pumped on to remove gases which boil out of the glass. A seal may be made by directing a soft flame evenly around the middle of the constricted section. To avoid sucking a hole in the glass, care is taken to move the flame off a section as it begins to collapse. After the tube has been collapsed in an even fashion, one is left with a small, nearly solid section, which may still contain a threadlike opening between the two halves. Next, a small, very hot flame is directed around the nearly solid section, and while the flame is on the glass, the tube is separated from the apparatus by pulling it away. Finally, the sealed end is annealed with care to avoid further collapse of the tube.

GENERAL REFERENCES

Barr, W. E., and V. J. Anhorn, 1959, *Scientific Glassblowing*, Instruments Publishing, Pittsburgh.

Heldman, J. D., 1946, *Techniques of Glass Manipulation in Scientific Research*, Prentice-Hall, Englewood Cliffs, N.J.

Laboratory Glass Blowing with Corning Glasses, 1961, Bulletin, B-72, Corning Glass Works, Corning, N. Y.

Moore, J. H., C.C. Davis, and M. A. Coplan, 1983, *Building Scientific Apparatus*, Addison-Wesley, Reading, Mass. Glassblowing is covered in Chapter 3.

Parr, L. M., and C. A. Hendley, 1956, *Laboratory Glassblowing*, G. Newnes, London.

Wheeler, E. L., 1958, *Scientific Glassblowing*, Interscience, New York, Available from E. L. Wheeler, P.O. Box 4833, Chico , CA 95927.

PLASTICS AND ELASTOMERS

Many recent improvements in apparatus have resulted from the availability of new high-polymer materials. These substances exhibit a wide range of physical and chemical properties. Therefore, it is frequently necessary to exercise care in the choice of the rubber or plastic to be used for a particular application.

Rubbers and plastics have a much more limited range of useful temperatures than do metals or glass. Most polymeric materials undergo a low-temperature transformation to a brittle glassy state which often limits their utility at reduced temperatures. In addition, plastics lose their strength at high temperatures. Small changes in formulation may significantly alter the usable temperature range of a polymer material, so the glass transition temperatures and maximum usable temperatures quoted in this appendix are only approximate.

Elastomers and, to a lesser extent, plastics tend to swell in the presence of solvents. Frequently, this swelling is accompanied by a loss in strength. In addition, some plastics are dissolved by certain solvents, and all-polymeric materials may be chemically destroyed. It is in the areas of solvent and chemical resistance that most problems are encountered in the laboratory.

The permeability of high-polymer materials to atmospheric gases and solvents is much greater than for metal or glass. This is often a significant factor when air-sensitive compounds are being handled or when a high vacuum is desired. It is evident from Table III.1 that permeabilities may vary greatly from one material to the next. The rate of permeation decreases with an increased cross section of the plastic or elastomer, because the amount of gas which will pass through a membrane of thickness d is given by the equation

$$q = QA \left(\frac{P_1 - P_2}{d}\right)t$$

259

where A is the membrane area, $P_1 - P_2$ is the difference in partial pressures of a particular gas on the two sides of the membrane, t is the time, and Q is the permeability.[1] Values are presented in Table III.1 for the permeability Q of various materials to atmospheric gases. In most cases, the permeability increases with temperature; and it may be decreased by greater cross-linking or greater crystallinity.

From the data in Table III.1 it may be seen that cellophane and cellulose acetate have very large permeabilities to water; also, the solubility of water in these materials is great, so they are clearly unsuited as moisture barriers. Of the elastomers listed, silicone rubber has the highest permeability to air.

III.1 PLASTICS: GENERAL PROPERTIES AND FABRICATION

Most plastics may be classified as *thermosetting* or *thermoplastic*. At the molecular level the former are characterized by a three-dimensional cross-linked network. These thermosetting plastics soften progressively as heat is applied, but they do not exhibit a true melting point, and often the maximum usable temperature is higher than for a comparable thermoplastic. The thermoplastics, which are composed of high-polymer chains without cross-links, exhibit reasonably well-defined softening ranges and are often more susceptible to solvents than similar thermosetting polymers.

The method of fabrication for plastic items is influenced by whether or not one is dealing with a thermoplastic or a thermosetting plastic. In general, the first group may be heat-sealed, heat-molded, and solvent-cemented, but these possibilities are not available for the thermosetting resins. Both groups generally can be machined.

Irradiation of thermoplastics leads to cross-linking and thus imparts the properties of a thermosetting plastic. Tubing which has been expanded after the radiation treatment has a "memory" of its former size and will shrink to its original size when heated. Shrinkable tubing is available in a number of different plastics.[2]

Heat-sealing is routinely used in the "bagging" operation for the disposal of radioactive and/or highly poisonous materials from a glove box. The general approach in heat-sealing is to heat the film to the melting temperature and press the molten surfaces together. To accomplish this, a tool with a heated Teflon-coated roller is available commercially, and an accessory heat-sealing tip is available for many electric soldering guns.

It is possible to weld substantial pieces of thermoplastic by the use of a hot-

---

[1]While reasonably accurate for the permeation of simple nonpolar gases, this equation is less suitable for water vapor, so the permeability data quoted for this substance are of qualitative significance only.

[2]Raychem Inc., Menlo Park, CA 94025.

gas torch (Fig. III.1).[3] These operate by electrically heating air or nitrogen; this hot gas is directed on the seam to be welded while a rod of the same plastic is fed into the molten seam (Fig. III.2). This procedure is effective for joining poly(vinylidene chloride) (Saran), rigid polyvinyl chloride, and polyolefins. The heat gun is also useful for joining metal or glass tubing to plastic tubing, such as polyethylene.[4] To construct a vacuum-tight joint, the glass or metal tubing is heated above the melting point of polyethylene, and the polyethylene tube is softened and slipped over it. The joint is heated with the hot-air gun, and the polyethylene is worked into the joint with a cold metal tool. A satisfactory end-to-end seal may be made with polyolefin and similar thermoplastic tubing. The ends of the two parts to be joined are evenly melted with the hot-air gun and then pressed together. While still molten, the welt of plastic at the seal is pinched together with pliers or tweezers, and the resulting thin, circular band of plastic is later trimmed away. Thermoplastic tubing may be sealed off by a simple crimp seal. This involves rotating the plastic tubing in the jet of hot air until an even molten ring results. This section is then crimped together with tweezers or pliers, and the two halves are cut apart at the seal. For small-diameter tubing, a "test tube end" seal may be achieved by drilling a shallow hole of the same diameter as the tube in a metal plate. This plate is then heated on a hot plate to the melting point of the plastic, and the end of the plastic tube is forced into it. A removable collar of glass or metal tubing may be slipped over the plastic tube to prevent bulging above the seal and a film of talc or silicone stopcock grease may be used to prevent sticking to the heated mold. A simple alternate method of joining and sealing thermoplastic tubing such as Kel-F or polypropylene, is illustrated in Fig. III.3. The plastic tubing (1) is forced together inside a glass tubing mold (2), which is gently heated by a cool flame (4), and a rod is inserted to shape the interior of the weld (3). This is an excellent procedure, since it produces clean, strong welds and the necessary equipment is readily available in the laboratory.[5]

To bend thermoplastic tubing, a broad area of the tubing is evenly heated in the hot-air stream and then bent. Just as with glass tubing, uneven heating leads to a sloppy bend. A much more satisfactory bend can be obtained by heating the tubing in hot water or air, bending it with a metal-tubing bender (Appendix IV), and, while it is still in the bender, quenching the tubing in a stream of cold water. Also, a flare, which is suitable for joints, may be constructed by heating the tube end in hot water or air, shaping the end with a flaring tool (Appendix IV), and quenching.

Sheets of thermoplastic material may be hot-formed into a variety of smooth shapes by heating the sheet in an oven to a temperature at which it becomes pliable. The sheet is then bent over a form. In the laboratory this method is

---

[3]Plastic welding torches are available from tool suppliers.

[4]When welding polyethylene and polypropylene, clean, freshly cut surfaces should be prepared, and nitrogen should be used in the torch because a thin film of oxide prevents a satisfactory bond.

[5]Neil Bartlett kindly supplied a description of this technique.

Table III.1. Permeability Q [in 10^{-10} cm²/(s·torr)] for Some Common Polymeric Materials[a]

Polymer	Temp °C	H_2	O_2	N_2	CO_2	H_2O
Cellulose and cellulose derivatives						
Cellophane (cellulose acetate)	25	...	0.0021	0.0032	0.0047	47–169,000
Cellulose acetate, plasticized	21	4	...	...	...	11,000–35,000
	25	...	0.78	0.28	2.38	
	30	...				
Cellulose nitrate	25	1.7	1.95	0.12	2.21	
Ethylcellulose, plasticized	30	...	26.5	8.4	41.0	
Elastomers						
Natural rubber, vulcanized	25	48	23	7.9	130	2,300
Butadiene rubber, vulcanized	25	42	19.2	6.5	139	
Buna S, vulcanized	25	40	17	6.2	122	4,400
Buna N, vulcanized	25	16	3.8	1.0	30	6,100
Neoprene, vulcanized (polychloroprene)	25	13	4.0	1.2	25.8	910
Butyl rubber, vulcanized	25	7.2	1.3	3.3	5.1	
Thiokol B, vulcanized	25	1.6	0.3	...	3.1	
Silicone rubber	30	...	91	(270)	460	65
Ethylene-propylene rubber	30	...	(air · · · · · · · · · 11)			
Fluorocarbon polymers:						
Teflon (polytetrafluoroethylene)	20	...	...	...	4.7	
Teflon FEP (tetrafluoroethylene hexafluoropropene copolymer)	25	...	4.5	1.9	10	

Polymer	Temp					
Kel-F, crystalline (polychlorotrifluoroethylene)	25	...	0.040	0.005	0.21	<0.3
Polyamides and polyesters:						
Nylon 6 [poly(6-aminocaproic acid)]	20	...	...	...	...	40–4,000
	25	...	...	...	0.088	
	30	...	0.038	0.0095	...	
Mylar [poly(ethylene terephthalate)]	30	...	0.045	0.011	0.15	
Polycarbonate poly(4,4′-isopropylidene diphenylene carbonate)	25	...	1.4	0.3	8.0	
Olefin polymers:						
Polyethylene, low density	25	(4)	...	...	...	21–66
	30	...	3.95	1.36	16.7	
Polyethylene, high density	30	...	0.51	0.18	2.1	
Polypropylene	30	...	2.3	0.44	9.2	
Polystyrene	25	(15)	...	...	...	920–1,300
	30	...	1.1	0.29	8.8	
Polyvinyl chloride	25	...	...	...	...	130–260
	30	...	0.3	0.11	1.5	
Saran; polyvinylidene chloride	30	...	5×10^{-3}	9×10^{-4}	0.03	
Pliofilm, plasticized; rubber hydrochloride	25	...	...	...	...	13–330
Acrylics:						
Poly(methyl methacrylate)	25	...	...	...	...	1,300

*These data are taken from H. Yashuda, "Polymer Handbook," J. Brandup and E. H. Immergut (eds.), V 13ff, Interscience Publishers, New York, 1966; G.J. van Amerongen, "Elastomers and Plastomers," R. Houwink (ed.), vol. 1, pp. 310ff, Elsevier Publishing Company, Amsterdam, 1950; R.P. Bringer, paper presented at Society of Aerospace Materials and Process Engineers, St. Louis, May 7-9, 1962 (fluorocarbon data, in part), Ethylene propylene rubber: "Nordel an Engineering Profile," E.I. du Pont de Nemours and Co., Elastomer Chemical Dept., Wilmington, DE., and Du Pont bulletin T-3B (Teflon FEP).

Fig. III.1. Plastic welding torch.

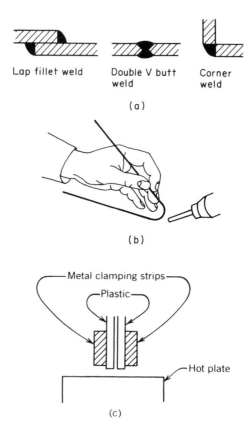

Lap fillet weld   Double V butt   Corner
                         weld   weld

(a)

(b)

(c)

Fig. III.2. Plastic welding practice. (*a*) Some typical welds. (*b*) Technique for feeding rod into weld. With the rod bent back into a 45° angle, it is easy to heat and the rod is forced into the seam. The torch should be fanned between the rod and the parts being welded, with somewhat greater concentration on the latter. (*c*) An edge seal accomplished on a hot plate.

suitable for the construction of simple, smooth shapes and is particularly successful with poly(methyl methacrylate) sheet stock.

As mentioned earlier, many thermoplastic materials dissolve or become tacky in the presence of a solvent, and this provides the basis of solvent cementing. The edges to be joined should be well mated; one of the edges is then soaked for about 5 min in the solvent. The part is removed from the solvent, and excess liquid is allowed to drip off. While still damp, the part and its mate are lightly

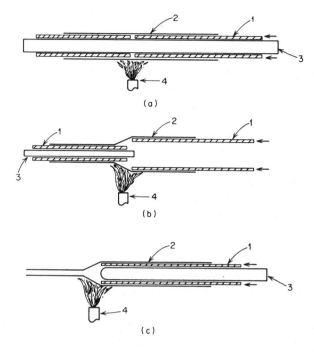

Fig. III.3.   Joining and sealing thermoplastic tubing. (*a*) End-to-end seal between tubes of identical diameter; (*b*) end-to-end seal between tubes of different diameter; (*c*) closed-end seal.

clamped together. When clear plastic is used, the joint should be transparent and nearly bubble free. Solvent sealing is especially useful in the fabrication of parts from cast poly(methyl methacrylate) sheet stock.

Most of the common machining operations can be performed on plastics which are not too soft or too brittle to withstand working. Plastics do not dissipate heat as effectively as metals, so care has to be taken to avoid overheating. In the case of Teflon, poisonous fumes may be given off by the excessively heated plastic; and many of the thermoplastics can be heated to the point that they flow and gum up. In general, overheating can be avoided by employing sharp drills and cutting tools and by appropriate adjustment of working rates. Heat dissipation can be increased by drilling with an in-out motion, and inert coolants, such as water or water-oil emulsions, may be used.

III.2 CELLULAR PLASTICS

At one time or another practically every plastic and elastomer has been processed in a cellular form. Only two of the most commonly encountered foamed plastics will be considered here: styrofoam and foamed polyurethane. Both of these materials have low thermal conductivity and are very useful as low-temperature insulators.

While unaffected by water, styrofoam is dissolved by many organic solvents and is unsuitable for high-temperature applications because its heat-distortion temperature is around 77°C. Molded styrofoam objects are produced commercially from expandable polystyrene beads, but this process does not appear attractive for laboratory applications because polyurethane foams are much easier to "foam in place." However, extruded polystyrene foam is available in slabs and boards which may be sawed, carved, or sanded into desired shapes and may be cemented. It is generally undesirable to join expanded polystyrene parts with cements that contain solvents which will dissolve the plastic and thus cause collapse of the cellular structure. This excludes from use a large number of cements which contain volatile aromatic hydrocarbons, ketones, or esters. Some suitable cements are room-temperature-vulcanizing silicone rubber (see below) and solvent-free epoxy cements. When a strong bond is not necessary, polyvinyl-acetate emulsion (Elmer's Glue-All) will work.

Urethane foams may be rigid, semirigid, or flexible depending on the ingredients and their proportions. The rigid urethane foams generally swell in the presence of organic solvents such as benzene, carbon tetrachloride, and acetone; however, exposure to these solvents and subsequent drying has little effect on the foamed plastic. Some of the flexible foams are more solvent sensitive, but this does not preclude the use of most cements. Flexible expanded urethane tubing is commercially available in a variety of sizes.[6] It provides a simple and efficient method for the insulation of tubing which carries refrigerants. For example, it has proven successful for the insulation of liquid nitrogen lines.

Foamed-in-place polyurethane is prepared by allowing a polyol [poly(ethylene glycol), polyester alcohols, etc.] to react with a diisocyanate in the presence of an amine catalyst. The gas which creates the foam may be a dissolved material, such as a Freon, which volatilizes during the exothermic polymerization reaction.[7] A second method involves the use of water in the reaction mixture; this hydrolyzes part of the isocyanate to produce an amine and CO_2 gas. The Freon-formed material is preferred for the insulation of low-temperature apparatus because the thermal conductivity of the foam is greatly reduced at low temperatures by the condensation of the Freon in the cells. It is probable that the long-term effectiveness of this phenomenon must be maintained by surrounding the foamed plastic with an airtight enclosure which will prevent diffusion of air into and Freon out of the cells.

---

[6]Polyurethane and polyvinyl foamed tubing for insulation are available from refrigeration supply houses.

[7]The ingredients for the preparation of polyurethane foam are available from many companies, including the Perlon Corp., Lyons, IL, and American Latex Products, Hawthorne, CA.

III.3 PROPERTIES OF SPECIFIC PLASTICS

A. Acrylics. This group of clear thermoplastic materials is frequently encountered under the trade name Plexiglas (Rohm and Haas). The most common member of the series, poly(methyl methacrylate), has a softening point of 124°C, at which temperature sheet stock, rods, and tubing may be formed into special shapes. In addition, the plastic may be machined; however, care has to be taken to avoid overheating. Cast poly(methyl methacrylate) sheet stock may be bonded by means of a solvent such as methylene chloride or 1,2-dichloroethane. Giauque and co-workers used a cement composed of equal volumes of methyl methacrylate monomer and dichloromethane. This is used exactly like the pure solvent cements. After the lightly clamped parts had dried for a few hours, the joint was further cured and annealed by heating the part to 50°C for about 1 day. It was found that a joint of this sort gave reliable vacuum-tight performance at liquid helium temperature.[8]

B. Chlorinated Hydrocarbon Plastics. Polyvinyl chloride is often fabricated into sheets and tubing. Some formulations are reasonably transparent. Generally, plasticizers are included in the formulation to increase the flexibility, and tubing of this sort is a familiar laboratory item under the Tygon trademark. A number of formulations are available, but one of these, Tygon R-3603, is recommended for general laboratory use. It makes a good seal with glass tubing and is claimed to have a vapor pressure of 10^{-7} mm at 21°C. The amount of gas desorbed by polyvinyl chloride is very low in comparison to rubbers. Furthermore, the data in Table III.1 show that polyvinyl chloride has a lower permeability to atmospheric gases than rubber, so heavy-walled Tygon tubing is to be preferred to rubber when a flexible vacuum connection is required. While this tubing is less susceptible to deterioration than most rubber tubing, it tends to stiffen and become milky in appearance after long exposure to water. It is essentially inert with aqueous acids and alkalies but is attacked by most organic solvents, as well as Cl_2, HCl, and BCl_3. The R-3603 formulation becomes brittle at −45°C and loses much of its strength around 93°C. Rigid polyvinyl chloride pipe is available from hardware and plumbing suppliers. It is easily cut with a saw and may be joined to elbows, T's, and other fittings by means of a special cement. The grade which is made to carry drinking water has a low content of volatiles and has been found suitable for rough- and medium-vacuum systems.[9]

Polyvinylidine chloride (Saran) is a tough plastic which is extensively used in laboratory drainpipes. This form of the plastic is rigid, but flexible transparent films are used widely in food packaging. Polyvinylidine chloride becomes brittle at 10°C and loses most of its strength at 77°C. It has unusually low permeability

[8]W. F. Giauque, T. A. Geballe, D. N. Lyon, and J. J. Fritz, *Rev. Sci. Instr.*, 23, 169 (1952).
[9]R. L. Hartman, *Rev. Sci. Instr.*, 38, 831 (1967).

to gases and is chemically attacked by only *very* strong oxidizing or reducing agents. The solvent resistance is fairly good, and therefore solvent bonding is not easy. Both edges to be bonded should soak in the solvent (preferably warm) for an extended period of time before they are joined. Cyclohexanone, *o*-dichlorobenzene, and dioxane are suitable for solvent bonding. Epoxies form string bonds to vinylidine chloride, and welding or heat-sealing is quite successful.

C. Epoxy Resins. The epoxy resins are thermosetting plastics which have great strength and the ability to form tenacious bonds with most surfaces. Furthermore, the cured resin is resistant to many solvents and chemicals. (Some epoxy resins are decomposed by acetic acid, and all are attacked by very strong oxidizing agents.) Because of this combination of properties, epoxy cements are frequently used to bond metal, glass, wood, and plastics.

Epoxy resins tend to outgas under vacuum, primarily through the slow desorption of moisture and low-molecular-weight organic material. However, good epoxy cements which are compounded without filler materials or solvents have acceptable outgassing rates for high-vacuum applications.[10] Another property which must be anticipated for some applications is the high thermal expansion coefficient of the cured resin. Because of this property, a glass tube which contains a solid plug of the cured resin will often break when heated or cooled.

The basic ingredient in the epoxy resin is a viscous polymer which generally contains two epoxide groups

$$\overset{\displaystyle O}{\overset{\displaystyle /\;\backslash}{(-O-CH_2-CH-CH-)}}$$

per molecule. The further polymerization of this resin to a hard infusible mass is brought about by mixing it with a curing agent (also called a hardener or a catalyst) just prior to application. The curing agents vary in composition and mode of action, but one common type involves tertiary amines which catalyze the attack of one epoxide group on the next to interlink the original resin through $-O-(CH_2-CH(R)-O)_n$ groups. Reactive curing agents, such as primary and secondary amines and alcohols, are also employed and are incorporated into the polymer during the cure. The proportions of resin to curing agent recommended by the manufacturer should be followed with reasonable accuracy to avoid an incompletely cured product or an excessively fast and exothermic cure which

[10]The Araldite resins (e.g., Araldite 1 and 101) manufactured by Ciba Products Corp., Summit, NJ, and the Epon Adhesives manufactured by Shell Chemical Co., Pittsburgh, PA, are widely used in vacuum applications. K. S. Balain and C. J. Bergeron, *Rev. Sci. Instr.*, 30, 1058 (1959), found that Stycast 2850 GT, made by Emmerson and Cumming Inc., Canton, MA, gave a seal which was suitable for vacuum and pressure application at low temperatures. W. R. Roach, J. C. Wheatley, and A. C. Mota de Victoria, *Rev. Sci. Instr.*, 35, 634 (1964) used Epipond 100-A (Furane Plastics, 4516 Brazil St., Los Angeles, CA) for the construction of vacuum-tight, low-temperature electrical-feed throughs. It should be noted that the opaque epoxy cements which are sold in hardware stores often contain fillers.

leaves no time for application. The uncured resin and hardener should be handled in a well-ventilated area, and contact with the skin should be avoided because the amines employed in the hardener are toxic skin irritants, and the resin may cause dermatitis or an asthmalike condition with sensitive individuals. "One-component" epoxy preparations are also available in which the catalyst is activated by the application of heat.[11]

D. Fluorocarbon Polymers. Four different fluorocarbons account for the bulk of the laboratory applications: polytetrafluoroethylene, Teflon PTFE; poly(chlorotrifluoroethylene), KEL-F; tetrafluoroethylene-hexafluoropropylene copolymer, Teflon FEP; and tetrafluoroethylene-perfluorovinyl ether copolymer, PFA. These polymers are inert with most chemicals and solvents at room temperature and exceptionally inert with oxidizing agents. They also have an exceptional resistance to temperature extremes. However, they are decomposed by liquid alkali metals, solutions of these metals in liquid ammonia, and carbanion reagents such as butyllithium. Teflon retains some of its compliance at liquid hydrogen temperature. The maximum temperature which is recommended for continuous service is 260°C for Teflon PTFE and PFA, and about 200°C for Kel-F and Teflon FEP.

Teflon is an opaque, flexible solid with a waxy feel. It has a self-lubricating quality which makes it a useful material for the packing in valves. However, it tends to cold-flow, and must be well supported on all sides if it is to be used under pressure. For this reason, Teflon O-rings have not been particularly successful in high-vacuum applications; but other types of gaskets which are supported better, such as simple flat gaskets between flanges, are serviceable. The flexibility and tendency of Teflon to cold-flow may lead threaded parts in Teflon to strip when they are under much force. Conventional Teflon PTFE can not be heat-molded or heat-sealed, and cementing is not very satisfactory. The principal method of fabrication in the shop is machining. Teflon FEP is similar in most respects to the conventional material; however, it is more transparent and may be readily heat molded and heat-sealed. Both types of Teflon tend to outgas when used in high-vacuum apparatus.

Kel-F is a somewhat more rigid plastic than the Teflons and has much less tendency to cold-flow. Therefore, it will bear a continuous load and is preferable to Teflon for threaded items. As with Teflon FEP, Kel-F may be heat-formed and sealed (330 to 400 °C). It is somewhat more susceptible to swelling than the Teflons when it is contacted with halogenated and aromatic hydrocarbons at elevated temperatures. Also, it is slightly more susceptible to oxidation at elevated temperature. However, at room temperature it gives satisfactory performance in the presence of solvents and strong fluorinating agents. Kel-F is useful in vacuum apparatus and is an excellent seat material for metal valves. Porous, sintered filter disks of Kel-F and Teflon are available commercially.[12] The

---

[11]Ciba Products Corp, Summit, NJ.

[12]Pall Trinity Micro Corp., Cortland, NY.

former may be heat-sealed into Kel-F apparatus. When heated in air, the toxic gases COF_2 and HF are evolved above 400°C for Teflon PTFE and above 350°C for Teflon FEP. When machining either of these polymers or heat-sealing the latter, care should be taken to avoid overheating, and good ventilation is advised.

E. Nylon. The most common member of this group of polyamide plastics is nylon 66 (polyhexamethylenediamine adipic polyamide), which has high strength and a very low tendency to cold-flow. It has a self-lubricating quality and is used for small bearings, gears, cams, and threaded parts. It has a greater tendency to outgas under vacuum than most plastics. It is easily machined and may be cemented with epoxy resins but not with solvents. It is dissolved by phenol and phenolic compounds and by concentrated acids. Nylon should not be used for long periods above 149°C in the air, and it softens rather sharply about 100°C above this temperature.

F. Polyester Resins. Polyester resins are made by the esterification of polyhydrolylic alcohols with polybasic carboxylic acids. High-molecular-weight polyesters with OH end groups are used in conjunction with isocyanates to produce many of the urethane plastics. When the polyester contains some unsaturated sites, polymerization may take place to cross-link the polyesters and produce a thermosetting plastic. This cure is initiated by a catalyst, such as an organic peroxide, and accelerators are also included to promote the action of the catalyst. Fiberglass-reinforced polyester items have high impact resistance and good chemical and thermal resistance. Therefore, this material is used in the construction of fume hoods, dry boxes, and tanks for chemicals. The ingredients for making polyester-reinforced items are available from boat shops.

Clear polyester sheet stock and rods are available[13] and, like reinforced polyester plastic, may be drilled, sawed, and machined. This clear plasic is harder than poly(methyl methacrylate) and much more solvent resistant than either poly(methyl methacrylate) or polystyrene. As with all thermosetting plastics, it may not be heat-formed or solvent-bonded. However, bonding with epoxy cements is satisfactory. This plastic is claimed to give continuous service at 80°C and intermittent service up to 150°C.

Transparent polyester films such as Mylar (Du Pont) are much stronger than most other plastic films. However, Mylar is not as resistant to oxidizing agents and bases as are polyolefin films.

G. Polyolefins. The thermoplastic hydrocarbons polyethylene and polypropylene are flexible, inexpensive materials with good chemical resistance. Two types of polyethylene are available: low density and high density. The former contains branched polymer chains which impart flexibility. The maximum usa-

---

[13]A series of materials of this nature is produced under the Homalite trade name by G-L Industries, Inc., The Homalite Corp. (a subsidiary), 11 Brookside Dr., Wilmington, DE.

ble temperature is about 80°C. High-density polyethylene contains linear chains and is more rigid, higher melting, less permeable, and more solvent resistant than the conventional material. The maximum usable temperature, about 120°C, is strongly dependent on the synthetic details. Polypropylene is similar to, but slightly more brittle than, high-density polyethylene. Polypropylene also is useful at higher temperatures than polyethylene.

All the polyolefins will swell in the presence of aromatic, aliphatic, and chlorinated hydrocarbons; this swelling increases with temperature, and at sufficiently high temperatures dissolution results. They are attacked by strong, warm oxidizing agents, but are resistant to strong reducing agents like sodium in liquid ammonia. Since they are not dissolved by solvents around room temperature, polyethylene and polypropylene cannot be solvent-bonded. Also, adhesives do not generally adhere well; however, as described earlier in this appendix, heat-sealing is very successful.

The aliphatic polyolefins are translucent to visible light and contain some strong absorptions in the medium-infrared, but they are transparent through most of the far-infrared and are therefore used as cell-window material for far-infrared spectroscopy. The use of low-density polyethylene in vacuum systems has been described by Duncan and Warren, who note that a $1/8$-in. wall thickness is sufficient to prevent collapse of tubing up to $1/2$-in. They found it relatively easy to evacuate their system to 10^{-6} torr after an outgassing period of several days.[14] In addition to the release of previously dissolved gases, the vacuum is impaired by the permeation of the plastic by atmospheric gases; however, these authors claim that this factor is not too serious since a vacuum of 10^{-4} torr could be maintained on their polyethylene system when it was left overnight without pumping.[15] Polyethylene is not desirable for low-temperature traps partly because of its low thermal conductivity and also because of its high brittleness temperature, which is about -50-$-120°$ and -37-$-120°C$ for low-density and high-density polyethylene, respectively. If a polyethylene item contains internal stress, it is liable to break below the brittleness temperature.

Polystyrene is an inexpensive transparent plastic which is often used in industry for the fabrication of parts by injection moulding. However, the tougher acrylic plastics are preferable for the construction of laboratory apparatus. Polystyrene is soluble in many organic liquids and, if strain free, may be solvent-bonded by the use of chlorinated hydrocarbons, benzene, or toluene. Special impact-resistant grades are available which are less susceptible to solvents and thus a little harder to solvent-bond than the conventional material. Polystyrene also may be welded.

---

[14]J. F. Duncan and D. T. Warren, *Brit. J. Appl. Phys.*, 5, 66 (1954).

[15]From calculations based on permeability data and some laboratory tests, the authors believe that the pressure rise quoted here is less than can be expected with low-density polyethylene. The permeability data in Table III.1 show that considerable improvement is possible by using high-density polyethylene.

III.4 ELASTOMERS

It will be noted from Table III.1 that the elastomers[16] are generally much more permeable to gases than are plastics. Also, there is a greater tendency for elastomers to swell in the presence of solvents. In addition to the basic polymer, most finished-rubber products contain fillers, such as carbon black, and curing and vulcanizing agents, such as sulfur. Other additives may be included to improve the flexibility (plasticizers) and decrease damage by oxidation. The sulfur which is used as a vulcanizing agent can lead to contamination of chemical systems. For this reason some sulfur-free rubber goods are available. It should also be noted that silicone rubber is generally vulcanized without the use of sulfur.

III.5 PROPERTIES OF SPECIFIC ELASTOMERS

A. Buna N (Nitrile Rubber). This rubber is a copolymer of butadiene and acrylonitrile. As the nitrile content is increased, the resistance to hydrocarbons increases, but the low-temperature flexibility decreases. In general, nitrile rubber swells more in the presence of polar solvents than nonpolar solvents and is not usually suitable in contact with ketones, nitro compounds, and halogenated hydrocarbons. It is resistant to dilute inorganic acids and alkalies. The brittleness temperature of formulations which contain low percentages of acrylonitrile (about 20%) is about $-55°C$, while 36 % acrylonitrile leads to an increase in brittleness temperature to $-27°C$. On the other hand, softeners may be included in the formulation which allow the brittleness temperature to be reduced to about $-54°C$ for the high-nitrile-content rubber. Use above 120°C is not recommended. Nitrile O-rings and gaskets are extensively employed for applications which require resistance to hydrocarbons. From the data in Table III.1 it will be noted that the permeability of Buna N to moisture is high, but Buna N O-rings are used with success on vacuum systems, and both static and dynamic seals are possible because this rubber has good resistance to abrasion and cold-flow.

B. Butyl Rubber. Butyl rubber is a copolymer of isobutylene with a small amount of isoprene. The outstanding characteristics of this rubber are its low permeability to gases, high strength, and resistance to oxidation. Because of the low permeability, butyl rubber O-rings are useful in high-vacuum service, and butyl rubber gloves are desirable for use in inert-atmosphere glove boxes. Butyl rubber tends to stiffen rapidly as the temperature is lowered so that it loses much of its resilience above the brittleness temperature, which is about $-45°C$. Use above about 100°C is not recommended. Butyl rubber swells excessively in con-

---

[16]Additional information on the properties of specific elastomers may be found in J. Brandup and E. H. Immergut, Eds., *Polymer Handbook*, 2nd ed., (New York: Wiley, 1975).

tact with aliphatic, aromatic, and chlorinated hydrocarbons. Its resistance to ketones is satisfactory, and contact with silicone oils and greases does not lead to excessive swelling. It is attacked by hydrogen sulfide, halogens, and bases but is reasonably resistant to acids.

C. Neoprene (Polychloroprene).

Neoprene is a polymer of 2-chloro-1,3-butadiene. It has good resistance to aliphatic hydrocarbons (not as good as Buna N, however). Much more swelling is noted with halogenated hydrocarbons, aromatic hydrocarbons, and carbon disulfide, but satisfactory performance is possible in the presence of these solvents. This rubber may be exposed to all concentrated acids except nitric without extensive damage. Neoprene swells excessively in the presence of ketones. The brittle temperature is a sensitive function of the amount of monomer other than chloroprene used in formulating the rubber; also, plasticizers are sometimes added to lower the brittle point, which is ordinarily about −40°C. The upper useful temperature is about 150°C; however, it is inadvisable to heat neoprene gaskets above 90°C, since they tend to take a permanent set around this temperature. Neoprene is still used as an O-ring and gasket material in high-vacuum systems, but other elastomers have taken its place in most vacuum and laboratory applications.

D. Ethylene-Propylene Rubber.

Ethylene–propylene rubber, as the name implies, is a copolymer of ethylene and propylene. This particular rubber has very good resistance to polar organic compounds like ketones, ethers, acetic acid, and amines. It also withstands silicone oils and greases, dilute acids, alkalies, and metal solutions in liquid ammonia. However, it swells considerably and loses its strength in aliphatic, aromatic, and chlorinated hydrocarbons. It is claimed to be serviceable from −54 to 149°C. This elastomer is very attractive for general laboratory O-ring ware. It also has good abrasion resistance and tensile strength, so ethylene propylene rubber O-rings are suitable for dynamic seals. The permeability of this rubber is fairly high (Table III.1).

E. Fluorocarbon Rubber.

The most familiar rubber in this series is Viton-A (DuPont), which is a copolymer of hexafluoropropylene and 1,1-difluoroethylene. This material is the best general elastomer for O-rings and gaskets to be used in high-vacuum work. It will withstand temperatures from −29 to 205°C; and it is resistant to acids and halogenated, aromatic, and aliphatic hydrocarbons, and many metalloid halides like BCl_3. However, it swells and breaks down in contact with oxygenated and basic solvents such as ketones, ethers, esters, amines, and ammonia. Acetic and chlorosulfonic acids also attack Viton-A. This rubber is not recommended for continuous use in aqueous basic media. It will be noted that Viton-A and ethylene-propylene rubber complement each other rather well, since the former is resistant to hydrocarbons and many oxidizing agents but attacked by certain polar organic solvents, while the latter rubber has the opposite characteristics.

A fluoroelastomer manufactured by Du Pont called Kalrez, has mechanical properties and resistance to oxidants which are similar to those of Viton. In contrast with Viton, Kalrez has good resistance to polar molecules such as amines, ethers, ketones, and esters. Kalrez is unique among the elastomers in its tolerance to both polar and nonpolar solvents. The cost of O-rings made from Kalrez is very high, but for certain critical applications this cost can be justified because of the outstanding range of solvent tolerance.

F. Natural Rubber. Natural rubber products have low resistance to hydrocarbons and relatively high permeability to gases and moisture. Natural-rubber goods are often more flexible than those made of synthetic rubber and have excellent tensile strength.

G. Silicone Rubber. Silicone rubber has outstanding flexibility at low temperatures coupled with good resistance to heat (about -95–$230°C$). Brief service up to $300°C$ is possible for most silicone rubbers. Plasticizers are not used with this rubber, which is a contributing factor to the very low weight loss of silicone rubber in high vaccum. These rubbers are generally resistant to nitric, acetic, and hydrochloric acid but are attacked by concentrated H_2SO_4, concentrated base, and HF. Resistance to aliphatic, aromatic, and chlorinated hydrocarbons is variable. The principal weaknesses of silicone rubber are high permeability to gases and poor resistance to tearing and abrasion.

Room-temperature-vulcanizing silicone rubber (General Electric and Dow Corning) is available at hardware stores and is very useful as an adhesive and sealant. Atmospheric moisture is necessary to effect the cure, so broad areas of impermeable materials should not be cemented with these preparations. The uncured material evolves acetic acid, and the cured material appears to lose some weight in high vacuum; but if used with moderation, it can be considered a satisfactory vacuum sealant for most chemical vacuum systems.

GENERAL REFERENCES

Data sheets on the solvent resistance of various plastics and rubbers are available from the primary manufacturers. In addition, a number of O-ring manufacturers publish information on the solvent and chemical resistance of elastomers used in O-rings. For example, the Parker Seal Company, 17325 Euclid Ave., Cleveland, OH, publishes a *Seal Compound Manual* which contains this information.

Diels, K., and R. Jueckel (H. Adam and J. Edwards, trans.), 1966, *Leybold Vacuum Handbook*, Pergamon, New York. This is an English translation of the second German edition. It contains information on the gas desorption and mechanical properties of plastics and elastomers.

Evans, V., 1966, *Plastics as Corrosion-resistant Materials*, Pergamon, New York.

Newmann, J. A., and F. J. Bockhoff, 1959, *Welding of Plastics*, Reinhold, New York.

SPI Plastics Engineering Handbook, 1976, 4th ed., Van Nostrand Reinhold, New York. This book is concerned with the commercial production of plastic items. However, it contains some information which is useful for the fabrication of plastic items in the laboratory or shop.

METALS

When metal apparatus is designed for chemical applications, it is usually necessary to consider corrosion and heat resistance as well as ease of fabrication. In this appendix these topics are discussed at a level which should help the experimentalist to specify the correct materials and construction procedures. No details of shop practice are given. The interested reader may pursue this topic in the general references cited at the end of this appendix, and in Chapter 10.

Except for a few very hard and very soft metals, intricate shapes may be machined from heavy metal stock. Parts with thin metal cross sections frequently can be constructed from tubing or sheet stock which may be bent and formed into a variety of shapes.

Permanent joints may be formed by soldering or welding. Solders melt below the melting temperature of the parts being joined, and must wet the surfaces of these parts if a bond is to be achieved. A flux is employed to promote this wetting process, and the residual flux must generally be removed before the parts may be used. For example, a borate flux is frequently employed for hard soldering (brazing and silver soldering). This flux forms an adherent glassy deposit which will slowly outgas in vacuum. Welding is performed by melting the parts to be joined at the seam and simultaneously melting a "filler" rod of the same or very similar material into the seam. Arc or gas welding may be used with most metals, but the least porous welds are obtained by inert-gas welding (Heliarc). In this process, an electric arc creates the elevated temperature, while a stream of helium or other inert gas blankets the molten area. When possible, inert-gas welds should be specified for vacuum apparatus. When the apparatus is large enough to be welded from the inside, this mode of welding should be specified because it avoids the presence of rifts and voids which may slowly outgas. Of necessity, small items must be welded from the outside, but this is often quite acceptable because the material is thin. When it is not thin, a bevel should be provided so that the seam is thoroughly penetrated. Some good and bad welding practices are illustrated in Fig. IV.1. Often an oxide film or scale is left by welding or soldering; this should be removed from parts which are used in high vacuum because metal oxides slowly outgas, because of the desorption of moisture.

The fabrication from seamless tubing is described in Section 10.1.

IV. 1 PROPERTIES OF SPECIFIC METALS

A. Aluminum. Aluminum (mp 660°C), is a light, soft metal with fair resistance to corrosion by laboratory fumes except hydrogen halide vapor. It has excellent electrical and thermal conductivity. Many aluminum alloys are used; some may be softened by heating to 550°C and quenching in water. The metal subsequently work-hardens or hardens upon standing for a few hours (that is, 24 ST and 14 ST alloys), and all aluminum alloys are easily machined. These alloys

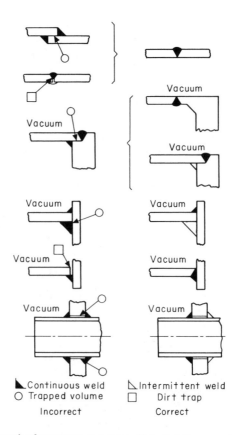

Fig. IV.1. Welding practice for vacuum apparatus. Note that the general approach is to weld on the inside and avoid dead spaces which may present leaks that are extremely hard to locate.

consist of more than 90% aluminum. Copper, magnesium, silicon, and transition metals are the most common alloying agents. Pure aluminum and most aluminum alloys may be welded, but the technique is not easy. Aluminum soft solders are generally unsatisfactory, but certain aluminum-containing eutectics make suitable hard solders. Electrodeposition on aluminum is not practical, but a tenacious oxide coating may be formed by the anodizing process, which involves electrolytic oxidation. The resulting coating may be dyed (e.g., black dyes are frequently used on parts used in spectroscopic work).

The chemical resistance of aluminum is fair, but any chemical which destroys the protective oxide coating will lead to rapid corrosion; therefore, strong acids, alkalies, and mercury must be kept out of contact with aluminum.

B. Copper. Copper (mp 1,083°C), is a moderately soft metal which has a high electrical and thermal conductivity. Electrolytic copper, 99.9% pure, is used for electrical wire; and free-cutting alloys, containing small amounts of sulfur or tellurium, are commonly used for sheet or bar stock. Copper may be machined, brazed, and soldered. However, because of its toughness, it is much less easy to machine than brass. It tends to work-harden but may be annealed by heating, followed by rapid quenching.

Copper is amalgamated by mercury and attacked by strong oxidizing agents, such as nitric acid, chlorine, and fluorine. However, the bulk metal forms a protective fluoride coating and is suitable for apparatus used with fluorine and oxidizing fluorides. The presence of ammonia promotes the air oxidation of copper, so copper and brass fittings should be avoided around NH_3. Copper is attacked by many sulfur-containing substances and is slowly attacked by most strong acids in the presence of oxygen.

Alloys of copper and zinc are called brass, while bronzes are composed of copper and tin. These alloys are generally easily machined, brazed, and soldered. Their chemical resistance is somewhat below that of copper. For example, brass is slowly attacked by fluorine and oxidizing fluorides. Brazed joints are much stronger than soft solder, and when used on well-matched surfaces, the brass tends to penetrate the joint and form a good, vacuum-tight seal. On the other hand, the highly fluid nature of molten brass makes it more difficult to fill gaping holes with brazing rod than with soft solder.

C. Gold. Gold (mp 1,063°C), is a soft, inert metal with good electrical and thermal conductivity. It is easily plated onto metals and vacuum evaporated onto various surfaces. Except for fused KOH and similar bases, strong oxidizing agents like aqua regia, and strong fluorinating agents, gold is chemically inert.

D. Indium. Indium (mp 157°C), is a soft metal which wets many surfaces and retains its ductility at very low temperatures. This combination of properties makes indium an excellent gasket material between dissimilar materials. For example, a gasket made of indium wire will give a vacuum-tight seal between quartz and metal from room temperature to below 4 K. Indium is also used in some special-purpose solders. It is attacked by cold dilute mineral acids, halogens, and other oxidizing agents, but is not affected by solutions of KOH.

E. Iron. Iron (mp 1,535°C), is softer and somewhat more corrosion resistant than the commonly encountered alloys. Cast iron contains a separate iron carbide phase as well as carbon dissolved in iron; in addition, other elements such as silicon, manganese, sulfur, and phosphorus are generally present. Gray iron and several other grades of cast iron contain graphitic carbon as a second phase dispersed thoughout the casting. The presence of graphitic phases leads to slow outgassing, which may be objectionable for certain exacting vacuum requirements. Iron castings may be somewhat porous because of rifts and voids in the granular structure. Such a casting is not suitable for vacuum work. Iron and steel are not amalgamated or appreciably dissolved by mercury; so iron or steel containers are used to transport mercury, and steel is used in some mercury-diffusion pumps.

Steel contains some carbon, about 2%, but considerably less than cast iron, which contains over 5%. Steel tends to become brittle at low temperatures, so it is not recommenced for high-pressure cylinders which will be cooled to liquid nitrogen temperature.

Slight variations in the constituents of iron or steel make them resistant to strong bases, nitric acid, and so on, but the judicious choice of irons and steels for specific chemical resistance is not practical in the laboratory.

Stainless steels are iron-based alloys which contain more than 12% chromium. A common composition contains 18% Cr and 8% Ni, and is designated as either 18:8 or type-304 stainless steel. Unlike ordinary carbon steel, the stainless steels in the 300 series do not become brittle at low temperatures. Stainless steel has a rather low thermal conductivity. It may be welded, brazed, or soldered and is machinable with some difficulty. Type 303 is the easiest to machine.

While the chemical resistance varies somewhat, stainless steel is fairly resistant to most acids and bases, is not amalgamated by mercury, and is generally resistant to oxidizing agents. While it can be used in fluorine handling, Monel and nickel are much better for this purpose. The resistance of stainless steel to atmospheric corrosion is an advantage in vacuum work because a corroded surface tends to outgas.

Certain alloys of iron, nickel, and cobalt (Kovar, Fernico, etc.) have thermal expansion curves which nearly match those of borosilicate glasses, and a good bond may be formed between the two. Kovar is similar to carbon steel in its chemical properties. For example, it oxidizes when heated in air and is not wet by mercury. It may be machined, welded, copper brazed, and soft soldered. Silver solders should not be used with Kovar since they may cause embrittlement. At low temperatures Kovar undergoes a phase transformation, and the change in expansion coefficient below this temperature may be sufficient to cause failure of a glass-to-Kovar seal. The transformation temperature usually is below

$-78.5°C$. The exact transformation temperature varies from one batch of the metal to the next and some samples of Kovar have transformation temperatures below $-196°C$. Another significant property of Kovar is its low thermal conductivity.

F. Lead. Lead (mp 328°C), is a soft metal which finds use in gaskets, soft solder, and radiation shields. While lead tank liners are used in some industrial processes, this metal is not attractive for laboratory containers.

G. Nickel. Nickel (mp 1,453°C), finds its primary use in the construction of apparatus to handle fluorine and volatile fluorides. In this situation the metal is rendered passive by a fluorine coating. Nickel plating is easily performed and provides a means of imparting corrosion resistance. The metal may be machined, silver soldered, copper brazed, or welded. However, the weld should be performed on clean surfaces because the presence of impurities containing lead, sulfur, phosphorus, and various low-melting metals leads to embrittlement and failure at the weld.

Monel is a type of alloy containing about 70% nickel, with the remainder mainly copper. Except for susceptibility to nitric acid, Monel is generally superior to stainless steel in corrosion resistance. Monel is similar to stainless steel in machinability and is welded like pure nickel.

H. Palladium. Palladium (mp 1,552°C), is soft and ductile but work-hardens. At elevated temperatures, the diffusion of hydrogen through palladium is rapid, which forms the basis of a method for the purification of hydrogen. The best performance is obtained from a palladium-silver alloy containing about 18% silver because, unlike pure Pd, this alloy does not undergo a phase transition in the presence of hydrogen. Palladium is not as inert as platinum and is attacked by sulfuric and nitric acids.

I. Platinum. Platinum (mp 1,769°C), is a malleable and ductile metal with high chemical resistance. Because of this resistance, it finds use in some laboratory ware. It is attacked by aqua regia, and, at somewhat elevated temperatures, by sulfur, phosphorus, halogens, and cyanide (in the presence of air). The coefficient of expansion for platinum nearly matches that of soft glass (flint glass), and an adequate seal between platinum wire and soft glass may be obtained.

J. Silver. Silver (mp 961°C), has the highest electrical and thermal conductivity of any metal. It is used in making electrical contacts and in the construction of some laboratory ware. It is easily electrodeposited, and both chemical and evaporation coating with silver are possible. Silver is generally less resistant to attack by oxidizing agents than is platinum or gold. However, it is resistant to fused alkalies and fused alkali-peroxide mixtures. Therefore, it is used to make crucibles for NaOH-Na$_2$O$_2$ fusions. At elevated temperatures oxygen diffuses through silver, and this property is occasionally used to purify oxygen.

Two types of silver solder are available. The soft-silver solders are generally alloys of silver with lead, which, like the common tin-lead soft solders, melt below 400°C. The soft-silver solders have the advantage over the Sn-Pb variety of greater resistance to creep under load. The hard-silver solders generally contain copper as the primary alloying constituent. These solders generally melt above 600°C. One common type of silver solder, AWP 355, is resistant to fluorine and many fluorides.

K. Tantalum. (Tantalum mp 2,996°C), is an extremely chemically resistant metal which is hard but ductile. Because of its high melting point and good corrosion resistance, tantalum is frequently used as a container for high-temperature melts. It is attacked by HF and other fluorides, as well as sulfur trioxide and nitrogen oxides.

L. Tungsten. Tungsten (mp 3,410°), has the highest melting point and lowest vapor pressure of the metals. The commonly available form of the metal contains trace impurities which impart brittleness. Tungsten wire is occasionally fibrous and therefore somewhat porous along the axis of the wire. Very pure forms have some ductility. Tungsten is oxidized in air but is otherwise fairly chemically inert. Mineral acids have little effect. The expansion coefficient is similar to Pyrex, so tungsten is frequently used in electrical lead-throughs for glass apparatus. It is not wet by mercury.

GENERAL REFERENCES

Hoke Corrosion Chart, Hoke Inc., 1 Tenakill Park, Cresskill, NJ. A compact tabulation for common metals and plastics.

Lee, J. A., 1950, *Materials of Construction for Chemical Process Industries*, McGraw-Hill, New York. Information on the corrosion resistance of materials arranged according to the corrosive agent.

Metals Handbook, 9th ed., Vols. 1–8, American Society for Metals, 1978–85.

Moore, J. H., C. C. Davis, and M. A. Coplan, 1983, *Building Scientific Apparatus*, Addison-Wesley, Reading, Mass.

Oberg, E., F. D. Jones, and H. L. Horton, 1984, *Machinery's Handbook*, 22nd ed., The Industrial Press, New York. A reference book on machine shop practice and properties of metals.

Treseder, R. S., 1980, *NACE Corrosion Engineer's Handbook*, National Association of Corrosion Engineers, Houston.

Welding Handbook, 5th ed., American Welding Society, New York, 1967. Contains useful sections on the principles of welding and welding specific metals like nickel and stainless steel.

VAPOR PRESSURES
OF PURE
SUBSTANCES

Vapor pressure data are important in the identification, separation, and general manipulation of known compounds on the vacuum line and for the characterization of new compounds. This appendix was prepared with these applications in mind. It is introduced with a discussion of the use and limitations of several analytical expressions for the representation of vapor pressure data. A description of a least-squares treatment of vapor pressure data for fitting to the Antoine equation is then presented. Finally, a table of vapor pressures at convenient slush bath temperatures is presented for 479 compounds.

V.1 ANALYTICAL REPRESENTATION OF VAPOR PRESSURE DATA

Most of the widely used vapor pressure equations have as their basis the Clapeyron equation

$$\frac{dP}{dT} = \frac{\Delta H}{T \, \Delta V} \tag{1}$$

which is an exact thermodynamic expression. ΔH is the molar heat of vaporization, and ΔV is the change in molar volume between the condensed and gaseous phases. If it is assumed that (1) the volume of the condensed phase is negligible by comparison to that of the gas, (2) the gas exhibits ideal behavior, and (3) the latent heat of vaporization is constant over the temperature interval of interest, Eq. (1) may be integrated to yield the simple relationship

$$\log P = - \frac{\Delta H}{2.3026RT} + B \tag{2}$$

or

$$\log P = \frac{-A}{T} + B \tag{3}$$

When converting to the absolute temperature scale T, it is important to recognize the convention $0°C = 273.15$ K. The constants A and B in Eq. (3) are determined empirically from experimental data by plotting $\log P$ versus $1/T$ and taking the slope as $-A$ and the intercept at $1/T = 0$ as B. Equation (2) is found to yield a fairly accurate fit to vapor pressure data for highly volatile substances and is often perfectly adequate for moderately accurate data below the boiling point.

The heat of vaporization varies slowly with temperature and eventually becomes zero at the critical temperature. In addition, the assumptions which were used to approximate ΔV (ideal gas behavior and negligible volume of the condensed phase) are inaccurate at high pressures. However, the assumptions are adequate for data of moderate precision (about $\pm 1\%$) in the 1–760-torr range, and in these cases the heat of vaporization may be determined from the A parameter as indicated in Eq. (3). Of course, different A and B parameters have to be determined for each particular phase (e.g., for the liquid, and for each solid phase, if more than one exists). If A and B are known for one phase and the heat of transition (fusion or solid-state transition) is known, the heat of vaporization and its equivalent, the A value for the second phase, may be calculated. The new value of B is then found by taking advantage of the fact that the vapor pressures for the two phases are identical at the transition temperature.

When accurate vapor pressure data are plotted as $\log P$ versus $1/T$ over an appreciable pressure range, it is usually found that there is some curvature to the plot. This defect in Eq. (3) is largely overcome by an empirical modification proposed by Antoine:

$$\log P = A - \frac{B}{t + C} \tag{4}$$

The introduction of a third constant C leads to an improved equation which has been tested in some detail and found to be the best three-parameter vapor pressure equation for general use.[1,2] Usually the temperature term in Eq. (4) is in degrees Celsius. Because of its simple form and generally good accuracy, this equation has been chosen for use in some of the most extensive compilations of

[1]G. W. Thomson, *Chem. Rev.*, 38, 1 (1946).
[2]M. Kh. Karupet'yants and Kuang-Yueh Ch'ung, *Chem. Abs.*, 53, 17613d (1959).

vapor pressure data (see below). Equation (4) is found to give a generally suitable fit between 30 torr and 3 atm, and it gives a much better fit between 50 and 760 torr. Thomson notes that the accuracy is poor when the reduced temperature exceeds 0.75 ($T_{red} = T/T_{crit}$). However, it is often much simpler to use a second Antoine equation above this temperature, adjusted so that the two coincide near $T_{red} = 0.75$, rather than fit the entire set of vapor pressure data with a single, more complex expression. The drawbacks of the equation in the high-pressure region are of little consequence in vacuum line work, where pressures greater than an atm are seldom attained. The deviations which occur below 10 torr are of more importance because it is often necessary to extrapolate data very far into the low-pressure region to estimate the temperature which must be used to trap a component effectively. Under these conditions, an extrapolation based on Eq. (3) is generally reliable and can be readily accomplished graphically.

The C value in Antoine's equation appears to be the least critical parameter, and some workers use a fixed value of 230 for all compounds. However, Thomson has found that there is a correlation between C and the boiling point of a substance. For elements which yield monatomic vapors, and for subtances which boil below $-150°C$ the correction leads to

$$C = 264 - 0.034\, t_b$$

where t_b is the normal boiling point in degrees Celsius. For other compounds he finds

$$C = 240 - 0.19\, t_b$$

The relationships are not exact, so if the data are good it is better to determine C directly.

Another rather simple semiempirical multiparameter equation has been used to fit vapor pressure data:

$$\log P = \frac{-A}{T} + BT + CT^2 + D \tag{5}$$

Multiplying through by T and rearranging leads to

$$T \log P = -A + DT + BT^2 + CT^3$$

The parameters A through D in this equation are easily fit by means of a least-squares routine. In several test cases it was found that the use of four parameters in Eq. (5) gives a slightly poorer fit to vapor pressure data than does the three-parameter Antoine equation.

It appears that the variation of the heat of vaporization with temperature is one of the primary sources of inaccuracy of Eq. (2), and the extra empirical parameters in Eq. (4) introduce a variation of ΔH with temperature. In a somewhat different approach, it is assumed that the heat of vaporization is a linear function of temperature, and the remaining assumptions involved in the derivation of Eq. (2) are retained. With these approximations the Clapeyron equation may be integrated to yield the Kirchoff equation

$$\log P = \frac{-A}{T} + B - C \log T \tag{6}$$

Nernst proposed a somewhat more elaborate equation, which may be written as

$$\log P = \frac{-A}{T} + B - C \log T + DT \tag{7}$$

He found that C frequently is close to 1.75, and it is the practice of some workers to use this value for all substances and vary only A, B and D. While Eqs. (6) and (7) give a better fit to the vapor pressures of some compounds than the Antoine equation, the accuracy is more often not quite as good. Also, the estimation of vapor pressures by the extrapolation of Eqs. (6) and (7) beyond the highest pressure of the original data appears to lead to more error than one would obtain upon extrapolation of the Antoine equation. Finally, Eqs. (6) and (7) are much more difficult to use when one wishes to find the temperature corresponding to a given pressure.

Many other vapor pressure equations have been used; the interested reader may refer to Partington's treatise for some of these.[3] One of the most successful equations for the faithful reproduction of vapor pressures from low pressures to the critical pressure was devised by Frost and Kalkwarf.[4]

$$\log P = A + \frac{B}{T} + C \log T + \frac{DP}{T^2}$$

This equation is appropriate for iterative computer calculation. Simplifications have been proposed by Thodos.[5]

V.2 LEAST-SQUARES FITTING PROCEDURE FOR THE ANTOINE EQUATION

When precise vapor pressure data are available, it is desirable to employ a least-squares fitting procedure to the Antoine equation. The formulation of this problem is available in detail in the literature and will be summarized here.[6]

[3]J. R. Partington, *An Advanced Treatise on Physical Chemistry*, Vol. 2 (London: Longmans, Green, 1951), pp. 265ff.

[4]A. A. Frost and D. R. Kalkwarf, *J. Chem. Phys.*, 21, 264 (1953).

[5]R. C. Reid and T. K. Sherwood, *The Properties of Gases and Liquids*, 2nd ed. (New York: Mc-Graw-Hill, 1966), Chap. 4. References and a discussion of the Frost-Kalkwarf equation may be found here. Also, a rather complete comparison of various reduced equations is presented. These are most useful at high pressures.

[6]C. B. Willingham et al., *J. Res. Nat. Bur. Std.*, 35, 219 (1945).

A. Derivation of Equations. Equation (4) may be rearranged to the form

$$(A - \log P)(C + t) - B = 0 \tag{8}$$

If the left side of Eq. (8) is multiplied out and a change of constants made, this equation becomes

$$at + b + c \log P - t \log P = 0 \tag{9}$$

where $a = A$, $b = (AC - B)$, and $C = -C$.

Since Eq. (9) is linear in the constants a, b, and c, data may fit to this equation using least squares. To reduce rounding errors, it is easy to make initial estimates of a, b, and c by using Eq. (9). These initial estimates, which will be called a_0, b_0, and c_0, may be calculated from three experimental points, preferably spread out over a substantial pressure range. This results in the generation of three simultaneous linear equations of the form of Eq. (9).

Substituting a_0, b_0, and c_0 into Eq. (9) gives

$$F_0 = a_0 t + b_0 + c_0 \log P - t \log P = 0 \tag{10}$$

which may be added to and subtracted from Eq. (9) to obtain

$$\alpha t + \beta + \gamma \log P + F_0 = 0 \tag{11}$$

where $\alpha = (a - a_0)$, $\beta = (b - b_0)$, $\gamma = (c - c_0)$, and F_0 is defined in Eq. (10).

Equation (11) is the form of the Antoine equation used to set up the normal equations for the least-squares calculation:

$$\begin{aligned}
\alpha\Sigma(t^i)^2 \quad &+ \beta\Sigma t^i \quad &+ \gamma\Sigma t^i \log P^i &= \Sigma f_0^i t^i \\
\alpha\Sigma t^i \quad &+ \beta(1) \quad &+ \gamma\Sigma \log P^i &= \Sigma f_0^i \\
\alpha\Sigma t^i \log P^i &+ \beta\Sigma\log P^i &+ \gamma\Sigma(\log P^i)^2 &= \Sigma f_0^i \log P^i
\end{aligned} \tag{12}$$

where $f_0^i = a_0 t^i + b_0 + c_0 \log P^i - t^i \log P^i$.

These equations are solved simultaneously for α, β, and γ, the resulting values used to find a, b, and c, and finally these values are used to find A, B, and C. The entire procedure is easily carried out on a hand-held calculator, as demonstrated in the example below. If many calculations need to be done, these calculations can be done on a computer.

B. Example Calculation: The Antoine Equation for Water. Table V.1 lists the vapor pressure data for water from 20 to 80°C. Values of a_0, b_0, and c_0 are calculated from the data at 20, 50, and 80°C by solving the following three simultaneous equations:

$$\begin{aligned}
a_0(20) + b_0 + c_0 \log(17.535) &= 24.878 \\
a_0(50) + b_0 + c_0 \log(92.51) &= 98.309 \\
a_0(80) + b_0 + c_0 \log(355.1) &= 204.028
\end{aligned}$$

$$a_0 = 8.07577$$
$$b_0 = 154.140$$
$$c_0 = -233.761$$

Next, the summations appearing in Eq. (12) are found for all of the data points. This leads to the set of three simultaneous equations:

$$\alpha(87,420) \; + \beta(1550) \quad + \gamma(3224.87) = -19.2937$$
$$\alpha(1550) \quad\;\; + \beta(1) \qquad + \gamma(60.1940) = -0.10986$$
$$\alpha(3224.87) + \beta(60.1940) + \gamma(121.536) = -0.51109$$

$$\alpha = -0.00307$$
$$\beta = 0.00002$$
$$\gamma = 0.07716$$

From this we obtain

$$a = 8.07270$$
$$b = 154.140$$
$$c = -233.684$$

Finally, reversing the first change of constants yields the Antoine constants

$$A = 8.07270$$
$$B = 1732.32$$
$$C = 233.684$$

and, therefore, the Antoine equation

$$\log P = 8.07270 - \frac{1732.32}{t + 233.684}$$

By comparing the second and third columns of Table V.1, it is seen that the agreement between the experimental and calculated water vapor pressure values is quite good.

V.3 CORRELATION AND ESTIMATION OF VAPOR PRESSURES

Occasionally one has available a single vapor pressure point—for example, the normal boiling point—but wishes approximate vapor pressures at other temperatures. Some of the simplest schemes are based on Eq. (2). If the compound is not likely to self-associate through hydrogen bonds or other specific interactions, the Trouton constant ($\Delta H_{vap}/T_{boiling}$) is approximately 21 cal/mol°. Substitution of this value into Eq. (2) shows that the vapor pressure is related to normal boiling point T_b by Eq. (13).

$$\log P_{atm} = 4.6 \left(1 - \frac{T_b}{T}\right) \tag{13}$$

Table V.1. Experimental and Calculated Vapor Pressure of Water

t	P_{exp}(torr)	P_{calc}(torr)
20	17.535	17.541
22	19.827	19.836
24	22.387	22.389
26	25.209	25.224
28	28.349	28.366
30	31.824	31.843
32	35.663	35.684
34	39.898	39.919
36	44.563	44.584
38	49.692	49.712
40	55.324	55.343
42	61.50	61.51
44	68.26	68.27
46	75.65	75.66
48	83.71	83.72
50	92.51	92.51
52	102.09	102.08
54	112.51	112.48
56	123.80	123.78
58	136.08	136.04
60	149.38	149.32
62	163.77	163.69
64	179.31	179.21
66	196.09	195.98
68	214.17	214.06
70	233.7	233.5
72	254.6	254.5
74	277.2	277.0
76	301.4	301.2
78	327.3	327.2
80	355.1	355.0

While very approximate, this relationship is useful for order-of-magnitude estimation of vapor pressures, which is often adequate in establishing trial conditions for trap-to-trap fractionation. For this purpose, a family of curves based on Eq. (13) is presented in Fig. V.1.

Another useful approximation is the Ramsay-Young rule, which states that for two related substances, the ratio of the temperatures at which they exert the same vapor pressure is constant.[7] That is,

[7]The nature of the approximation involved here is described by S. Glasstone, *Textbook of Physical Chemistry*, 2nd ed. (Princeton, NJ: Van Nostrand), p. 454.

$$\left(\frac{T_A}{T_B}\right)P = \left(\frac{T_A{}'}{T_B}\right)P'$$

Vapor pressure data often can be found for a substance similar to the one for which a single vapor pressure is available, and the tabulation by Stull (see below) is handy for this purpose, since the data are given at a series of fixed pressures.

A better estimate of vapor pressures for a variety of organic compounds can be made by means of the correlations of Hass and Newton. This procedure is easiest to use if the normal boiling point is known, but it may be used in an iterative fashion if the vapor pressure is known at an arbitrary temperature. The description is available in the *Handbook of Chemistry and Physics*, so it will not be repeated here.[8]

Finally, we note a simple but crude rule which is useful for quick estimates: For every 10°C rise in temperature the vapor pressure approximately doubles.

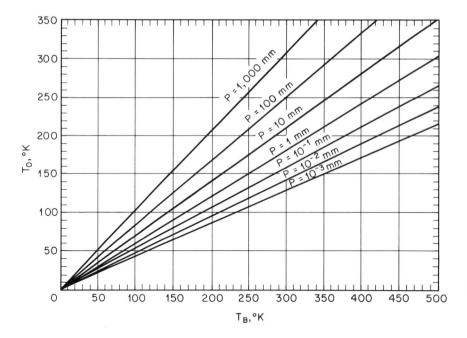

Fig. V.1. Approximate relationships between the vapor pressure at an arbitrary temperature T and the boiling point T_B.

[8]H. B. Hass and R. F. Newton, *Handbook of Chemistry and Physics*, 64th ed., (Cleveland, Ohio: Chemical Rubber Publishing Company, 1983–84), p. D-189. This information is presented in earlier editions and is indexed under: "boiling points, correction to standard pressure."

V.4 TABLE OF VAPOR PRESSURES

The following table of vapor pressure data (pages 290–311) is supplied as an aid to the planning of experiments and identification of compounds. When vapor pressure data are employed as a criterion of purity, the investigator may wish to evaluate the original work critically and obtain a precise interpolation of this data to the temperature of interest; therefore, references are included at the end of the table (pages 312–313). Most of the vapor pressures tabulated here were calculated on a digital computer using parameters from the references cited. In these calculations 0°C = 273.15 K. Since many of the original reports may have employed slightly different values for the ice temperature, a small error is sometimes introduced. Calculated values which knowingly represent an extrapolation or are otherwise suspect are enclosed in parentheses. When an analytical expression was not available or the given analytical expression was obviously erroneous, the original data were fit to Eq. (9) by least squares. The slush baths which correspond to the given temperature are methylcyclohexane, −126.59°; carbon disulfide, −111.95°; ethyl acetate, −83.6°; chlorobenzene, −45.2°; carbon tetrachloride, −22.9°; ice water, 0°C. All temperatures given in the table are in degrees Celsius and pressures are torr. A subscript s means the condensed phase is a solid.

Table of vapor pressures

Compound	Temperature, °C							$bp(°C)$	$mp(°C)$	Ref.
	-126.59	-111.95	-83.6	-45.21	-22.75	0.00	25.00			
	Vapor pressure, torr									
Aluminum:										
Aluminum borohydride $Al(BH_4)_3$	......	......	......	8.7	36	120	362	45.9	−64.5	1
Trimethylaluminum $(CH_3)_3Al$	......	......	......	......	......	......	12	124.7	15	LB
Antimony:										
Antimony pentachloride $SbCl_5$	......	......	......	......	......	......	1.2	......	2.8	LB
Antimony pentafluoride (SbF_5)	......	......	......	......	......	......	4.3	142.6	8.3	LB
Triethylstibine $(C_2H_5)_3Sb$	......	......	......	......	......	......	3.8	161.4	−98	2
Trimethylstibine $(CH_3)_3Sb$	......	......	......	(2.3)	(10)	35	109	80.6	−62.0	2
Stibine SbH_3	......	1.6	21	226	629	......	......	(−18.4)	−88	3
Arsenic:										
Arsenic pentafluoride AsF_5	......	3_s	106_s	......	......	......	......	−52.8	−79.8	LB
Arsenic trichloride $AsCl_3$	......	......	......	......	......	1.6	10	131.3	−16	LB
Arsenic trifluoride AsF_3	......	......	......	......	......	49	185	62.2	−5.9	LB
Arsine AsH_3	......	35	253	......	......	......	......	−62.3	−116.9	4
Dimethylarsine $(CH_3)_2AsH$	......	......	......	16	59	178	492	37.1	......	LB
Methyldichloroarsine CH_3AsCl_2	......	......	......	......	......	2.2	10	134.5	......	LB
Methyldifluoroarsine CH_3AsF_2	......	......	......	......	5.1	22	87	76.1	......	LB
Trimethoxyarsine $As(OCH_3)_3$	......	......	......	......	......	2.0	10	......	......	LB
Trimethylarsine $(CH_3)_3As$	......	......	......	7.2	30	98	295	50.4	......	LB
tris(trifluoromethyl)arsine $(CF_3)_3As$	......	......	......	15	58	187	551	(33.3)	......	5
Beryllium:										
Beryllium borohydride $Be(BH_4)_2$	......	......	......	......	......	......	8.0_s	91.3_s	123	6
Diethylberyllium $(C_2H_5)_2Be$	......	......	......	......	......	......	1.6	194	......	LB
Bismuth:										
Trimethylbismuthine $(CH_3)_3Bi$	......	......	......	......	2.6	10	37	107.1	−85.8	2

Boron:

Compound									Ref.
1,2-Bis(dichloroboryl)ethane $(BCl_2)_2C_2H_4$							(142)	−28.5	7
1,2-Bis(dichloroboryl)ethene $C_2H_2(BCl_2)_2$							(144)	−130	8
1,2-Bis(difluoroboryl)ethane $C_2H_4(BF_2)_2$					1.3	6.1	(35)	−31.5	8
1,2-Bis(difluoroboryl)ethene $(BF_2)_2C_2H_2$			9.9	44	2.3	9.4	15	−82	8
Bis(dimethylamino)borane $BH(N(CH_3)_2)_2$			25	109	156	507	105.7		9
Bis(dimethylamino)chloroborane $((CH_3)_2N)_2BCl$					380		146		LB
1,2-Bis(dimethylboryl)ethane $(B(CH_3)_2)_2C_2H_4$				1.7	8.5	35	98		10
Bis(perfluorovinyl)chloroborane $(C_2F_3)_2BCl$				2.8	1.2	5.8	100.5	−57.5	11
Borane carbonyl BH_3CO	5.7	25	231	2.7	12	44	−64.0	−137.0	LB
Borazine $B_3N_3H_6$			6.0	25	85	42	53.5	−58	12
Boron tribromide BBr_3				4.3	19	262	90.9	−46	LB
Boron trichloride BCl_3		2.4	50	175	476	69	12.4	−107	LB
Boron trifluoride BF_3	75	301					−99.9	−127	13
Bromodiborane B_2H_5Br		2.5		141	402		16.0	−104	LB
Diborane(6) B_2H_6	69	225	41	11			−93	−165.5	14
Diboron tetrachloride B_2Cl_4			2.3		44	150	(65.5)	−92.6	7
Diboron tetrafluoride B_2F_4		11_s	378				−34	−56	29
Dihydroxymethylborane $CH_3B(OH)_2$						2.5	ca. 100		20
Dimethylamine-borane $(CH_3)_2NHBH_3$						0.1		36	16
Dimethylaminodiborane $B_2H_5N(CH_3)_2$			6.6	29	101	304	50.3	−47.5	17

Table of vapor pressures (Continued)

Compound	Temperature, °C							bp(°C)	mp(°C)	Ref.
	−126.59	−111.95	−83.6	−45.21	−22.75	0.00	25.00			
	Vapor pressure, torr									
Dimethylaminodichloroborane (CH₃)₂NBCl₂	......	......	......	......	(1.4)	(7.4)	(26)	(112.3)	−43	LB
Dimethylarsine-borane (CH₃)₂AsHBH₃	......	......	......	......	1.4	8.1	40	(85.5)	......	LB
B,B′-Dimethylborazine B₃H(CH₃)₂N₃H₃	......	......	......	......	1.3	6.4	27	(107)	−48	18
N,N′-Dimethylborazine B₃H₃N₃(CH₃)₂	......	......	......	......	2.3	9.5	35	(108)	......	19
Dimethyldiborane (CH₃)₂B₂H₄	......	......	8.7	110	331	838	......	−2.6	−150.2	LB
Dimethylfluoroborane B(CH₃)₂F	1.2	6.0	66	657	......	......	......	−42.2	−147	20
Dimethylphosphine-borane (CH₃)₂PHBH₃	......	......	......	......	......	......	1.8	(174)	......	LB
Dimethylphosphine-dimethylborane (CH₃)₂PHB(CH₃)₂	......	......	......	......	......	......	3.9	......	......	LB
Dimethylphosphine-trimethylborane (CH₃)₂PHB(CH₃)₃	......	......	......	......	......	5.8	36	......	......	LB
Dimethylpropenylborane C₃H₅(CH₃)₂B	......	......	......	(4.9)	(22)	(76)	(242)	......	......	21
1,3,2-Dioxaborolane (CH₂O)₂BH	......	......	......	......	2.7	17	(91)	......	......	22
1,3,2-Dithiaborolane (CH₂S)₂BH	......	......	......	......	1.5	......	4.1	......	......	22
Hexaborane(10) B₆H₁₀	......	......	......	......	......	7.5	31	108	−62.3	23
Methoxydimethylborane (CH₃)₂BOCH₃	......	......	1.2	25	96	(302)	......	21	......	LB
N-Methylborazine B₃H₃N₃H₂CH₃	......	......	......	1.4	6.7	25	84	(84)	......	19

Compound										Ref.
Methyldifluoroborane BCH_3F_2	3.5	18	202	(8.1)	(34)	(113)	(344)	(−63.2)	−130.5	20
Methyldivinylborane $(C_2H_3)_2CH_3B$										21
Methylphosphine-borane $CH_3PH_2BH_3$							(2)	158		15
B-Monomethylborazine $B_3H_2CH_3N_3H_3$							70	(87)	−59	18
Pentaborane(9) B_5H_9				3.7	4.9	20	209	(57)	−46.8	25
Pentaborane(11) B_5H_{11}				3.1	18	65	170	(63)	−122	26, 24
Perfluorovinyldichloroborane $C_2F_3BCl_2$				6.1	14	52	309	48	−108	11
Perfluorovinyldifluoroborane $C_2F_3BF_2$				142	27	97		−14	−96	11
Phosphorustrifluoride-borane PF_3BH_3	23	211	8.3	38	501	388	(219)	(−62.2)	−116.1	27
Tetraborane(10) B_4H_{10}				4.7	134	71	2.2[s]	18	−120	15
Tetraborane(8) carbonyl B_4H_8CO					21	1.5	7.5	(59.6)	−114.5	28
Tetraboron tetrachloride B_4Cl_4								(140)[s]		LB
Tetramethoxydiboron $B_2(OCH_3)_4$				3.0	13	47	151	(130)	−26	30
Tetramethyldiborane $(CH_3)_4B_2H_2$						(3.0)	14	68.6	−72.5	LB
Triethoxyborane $B(OC_2H_5)_3$								118	−92.9	31
Triethylborane $B(C_2H_5)_3$				(5.9)	1.5	10	49	102	−46.4	32
B,B′,B″-Triethylborazine $B_3(C_2H_5)_3N_3H_3$							3.0	(173)		33
Trimethoxyborane $B(OCH_3)_3$					9.6		146	66.3		15
Trimethylamine-borane $(CH_3)_3NBH_3$						40	0.9	94		16
Trimethylarsine-borane $(CH_3)_3AsBH_3$							(1.5)[s]	154		LB
Trimethylborane $B(CH_3)_3$	1.3	20	20	245	683			−20.1	−161.5	32
B,B′,B″-Trimethylborazine $B_3(CH_3)_3N_3H_3$							ca. 10[s]	(129)	31.5	19
N,N′,N″-Trimethylborazine $B_3H_3N_3(CH_3)_3$						2.9	12	(134)	−8	19

Table of vapor pressures (Continued)

Compound	Temperature, °C (Vapor pressure, torr)							bp(°C)	mp(°C)	Ref.
	−126.59	−111.95	−83.6	−45.21	−22.75	0.00	25.00			
Trimethylboroxine B₃(CH₃)₃O₃	……	……	……	……	(4.8)	21	80	(79)	……	20
Trimethyldiborane (CH₃)₃B₂H₃	……	……	……	9.6	38	122	357	45.5	−122.9	LB
Tris(dimethylamino)borane B(N(CH₃)₂)₃	……	……	……	……	……	……	4.3	152	−16	34
Tris(perfluorovinyl)borane (C₂F₃)₃B	……	……	……	……	……	5.0	23	(104.5)	−107	11
Trivinylborane (C₂H₃)₃B	……	……	……	(2.9)	(15)	(64)	(235)	(55)	……	21
Cadmium:										
Dimethylcadmium (CH₃)₂Cd	……	……	……	……	……	9.8	36	105.7	−4.2	2
One carbon:										
Bromomethane CH₃Br	3.9_s	……	3.8	71	242	661	……	3.6	−93.6	MCA
Carbon dioxide CO₂	……	27_s	495_s	……	……	……	……	-78.5_s	……	LB
Carbon tetrabromide CBr₄	……	……	……	……	……	……	$(0.8)_s$	191.9	90.1	LB
Carbon tetrachloride CCl₄	……	……	……	8.3	33	113	76.7	−22.75	LB
Carbon tetrafluoride CF₄	……	……	……	……	……	……	……	−128	−183.7	LB
Carbonyl chloride COCl₂	……	……	(2.6)	56	200	563	……	7.6	−127.8	D
Carbonyl fluoride COF₂	37_s	124	751	……	……	……	……	−83.4	−114	LB
Chlorodifluoromethane CHClF₂	……	(4.6)	(62)	617	……	……	……	−40.8	−160	35
Chloroform CHCl₃	……	……	……	(3.1)	16	60	195	61.7	−63.49	D
Chloromethane CH₃Cl	……	……	23	279	……	……	……	−24.22	−97.7	MCA
Chlorotrifluoromethane CF₃Cl	31	107	667	……	……	……	……	−81.2	−181.2	LB
Dibromomethane CH₂Br₂	……	……	……	……	2.2	11	45	96.95	−52.6	MCA
Dichlorodifluoromethane CF₂Cl₂	……	2.7	36	376	……	……	……	−29.9	−160	LB
Dichlorofluoromethane CHCl₂F	……	……	(2.2)	(51)	185	531	……	8.9	−135	35
Dichloromethane CH₂Cl₂	……	……	……	8.2	40	(138)	(435)	39.8	−95.1	MCA

Compound	1	2	3	4	5	6	7	b.p. (°C)	m.p. (°C)	Ref.
Difluoromethane CH_2F_2	2.6	13	126					−51.6	.....	MCA
Fluoromethane CH_3F	19	76	563					−78.4	−141.8	MCA
Formaldehyde CH_2O			14	207	646			−19.1	−92	MCA
Formic acid $HCOOH$					42	140	42.8	100.6	8.4	MCA
Iodomethane CH_3I				10			406	42.43	−66.5	MCA
Methane CH_4			(10 mm at −196.5)					−161.5	−182.5	D
Methanol (CH_3OH)					5.7	29	125	64.7	−97.7	MCA
Trichlorofluoromethane CCl_3F			(1.5)	(27)	103	302		23.77	−111	35
Trifluoroiodomethane CF_3I		1.9	24	260	753			−22.5	.....	37
Trifluoromethane CHF_3	33	114	711					−82.4	−147	LB
Trifluoromethyl hypofluorite CF_3OF	75	251						−95.0	< −215	38
Two carbon:										
Acetaldehyde CH_3CHO			1.1	28	109	331		20.4	−123	MCA
Acetic acid CH_3COOH		67.8s	761.8s				15	117.9	16.7	MCA
Acetylene C_2H_2	12.8s							−84.8s	−80.8	39
Bromoethane C_2H_5Br			(2.6)	(12)	(50)	(163)	(469)	38.4	−118.6	MCA
Bromoethylene C_2H_3Br			13	(47)	157	424		15.8	−139.5	D
1-Chloro-1,1-difluoroethane $C_2H_3ClF_2$	1.1		13	152	444	(468)		−9.8	.....	40
Chloroethane C_2H_5Cl			(2.2)	(46)	(164)	521		12.27	−136.4	MCA
Chloroethylene C_2H_3Cl			14	181				−13.37	−153.8	D
Chloropentafluoroethane C_2ClF_5			59	574				−39.2	−99.4	LB
Chlorotrifluoroethane $C_2H_2ClF_3$	2.8		31	331				−27.9	−158	41
1,1-Dibromoethane $C_2H_4Br_2$				28			26	108.0	−63	MCA
1,2-Dibromoethane $C_2H_4Br_2$				3.8	71		12	131.4	9.79	D
1,2-Dichloro-1,2-difluoroethylene $C_2F_2Cl_2$			(7.1)	(87)	266	680		−112	.....	LB
1,1-Dichloroethane $C_2H_4Cl_2$			1.1	28	108	330	20.6	57.28	−97	MCA
1,2-Dichloroethane $C_2H_4Cl_2$				3.8	19	71	(4.9)	83.5	−35.7	D
1,1-Dichlorotetrafluoroethane CF_3CFCl_2					21		79	3	.....	40

Table of vapor pressures (Continued)

Compound	Temperature, °C — Vapor pressure, torr							bp(°C)	mp(°C)	Ref.
	−126.59	−111.95	−83.6	−45.21	−22.75	0.00	25.00			
1,2-Dichlorotetrafluoroethane $C_2Cl_2F_4$			7.1	87	266	680		3.0	−94	35
1,1-Difluoroethane $C_2H_4F_2$		1.4	23	283				−24.7	−117	MCA
1,1-Difluoroethylene CH_2CF_2	(48)	152						−85.7	......	40
Dimethyl ether C_2H_6O		1.6	23	283				−24.8	−138.5	MCA
Ethane C_2H_6	54	176						−88.6	−183.3	D
Ethanol C_2H_5OH					2.0	12	60	78.29	−114.1	MCA
Ethylene C_2H_4	156	459						−103.8	−169.2	77
Ethylene oxide C_2H_4O				44	172	495		10.7	−111	LB
Fluoroethane C_2H_5F		4.1	53	534				−37.7	......	MCA
Hexafluoroethane C_2F_6	22s	93s	570					−78.3	−100	LB
Iodoethane C_2H_5I	15			2.0	11	41	136	72.3	−111.1	MCA
Tetrafluoroethylene C_2F_4		62	483					−76.3	−142.5	36
1,1,2-Trichlorotrifluoroethane $C_2Cl_3F_3$					33	111	334	47.57	−35	35
1,1,1-Trifluoroethane CH_3CF_3	(2.8)	(12)	106					−47.6	......	40
Three carbon:										
Acetone C_3H_6O				3.4	18	67	(222)	56.5	−94.5	LB
Cyclopropane C_3H_6		3.0	42	437				−33.1	−126.6	LB
Methyl acetate $C_3H_6O_2$				3.0	16	61	211	57.8	−98.7	LB
Methyl acetylene C_3H_4			18	260				−23.2	−102.7	D
Methyl ethyl ether C_3H_8O				33	168	554		7.4	......	MCA
Propadiene C_3H_4		(1.9)	(45)	481				−34.5	−136	D
n-Propanal C_2H_5CHO				5.8	28	102	318	48.0	−80	MCA
Propane C_3H_8	(1.3)	(7.1)	76	660				−42.1	−187.7	D

Compound						b.p.	m.p.	Ref.
1-Propanol C_3H_7OH	....	....	....	3.3	21	97.2	-126.2	MCA
2-Propanol C_3H_7OH	....	....	1.2	8.3	45	82.6	-88.5	MCA
2-Propenal C_3H_4O	(2.0)	(6.6)	(27)	(89)	(269)	52.5	-87.7	LB
Propene C_3H_6	104	10	....	....	....	-47.7	-185.25	D
Four Carbon:								
1,2-Butadiene C_4H_6	(2.9)	52	178	496	....	10.85	-136.19	D
1,3-Butadiene C_4H_6	(7.3)	111	350	....	....	-4.41	-108.9	D
n-Butanal C_3H_7CHO	....	....	7.3	31	112	74.8	-96.4	MCA
n-Butane C_4H_{10}	(6.1)	94	299	....	....	-0.5	-138.4	D
2-Butanone C_4H_8O	1.4	....	7.1	30	100	79.6	-85.9	LB
1-Butene C_4H_8	(8.3)	123	380	658	....	-6.26	-185.4	D
cis-2-Butene C_4H_8	(4.1)	73	244	460	....	3.72	-138.9	D
Cyclobutane C_4H_8	2.5s	50	167	252	....	12.9	-50	LB
Dimethylacetylene C_4H_6	....	....	80	252	707	26.97	-32.26	D
Diethyl ether $(C_2H_5)_2O$	13	57	185	534	....	34.6	-116.3	MCA
1,2-Dimethoxyethane $C_4H_{10}O_2$	....	....	....	17.4	65	85.2	-69	42
1,4-Dioxane $C_4H_8O_2$	....	....	....	....	39	101.1	10	LB
Divinyl ether C_4H_6O	(17)	(72)	(234)	(672)	....	27.8	....	LB
Furan C_4H_4O	19	70	215	604	....	31.1	-85.6	LB
Isobutane C_4H_{10}	164	484	....	....	....	-11.73	-159.6	D
2-Methylpropene C_4H_8	126	376	391	....	....	-6.9	-140.4	MCA
Methyl n-propyl ether $C_4H_{10}O$	13	(34)	(145)	(457)	....	39.1	....	LB
Octafluorocyclobutane C_4F_8	(8.7)	101s	....	....	....	-6.6	-40.2	43
Tetrahydrofuran C_4H_8O	4.0s	12	48	163	....	ca. 63	....	D
trans-2-Butene C_4H_8	(5.3)	87	280	734	....	0.88	-105.6	D
Five carbon:								
Cyclopentane C_5H_{10}	....	(7.0)	31	107	317	49.26	-93.88	D
Cyclopentene C_5H_8	....	(9.1)	39	130	380	44.2	-135.1	D
2,2-Dimethylpropane C_5H_{12} (neopentane)	....	....	....	....	(532)	9.5	-16.55	D

Table of vapor pressures (Continued)

Compound	Temperature, °C							bp(°C)	mp(°C)	Ref.
	-126.59	-111.95	-83.6	-45.21	-22.75	0.00	25.00			
	Vapor pressure, torr									
2-Methyl-1,3-butadiene C₅H₈ (isoprene)	.....	.....	.....	(16)	(63)	(198)	(550)	34.07	-145.95	D
2-Methylbutane C₅H₁₂	.....	.....	.....	23	87	259	688	27.9	-159.9	D
n-Pentane C₅H₁₂	.....	.....	.....	14	58	183	512	36.07	-129.72	D
Six carbon:										
Benzene (C₆H₆)	.....	.....	.....	.....	3.9ₛ	24ₛ	95	80.1	5.53	D, J
Chlorobenzene C₆H₅Cl	.....	.....	.....	.....	.....	(2.5)	(12)	131.7	-45.21	7
Cyclohexane C₆H₁₂	.....	.....	.....	.....	.....	.....	98	80.7	6.55	D
Cyclohexanone C₆H₁₀O	.....	.....	.....	.....	.....	.....	4.6	155.6	-45	LB
Cyclohexene C₆H₁₀	.....	.....	.....	(1.0)	(6.0)	25	89	82.98	-103.5	D
Diethyleneglycol dimethyl ether C₆H₁₄O₃	.....	.....	.....	.....	.....	.....	(4)	162	-64	42
Di-n-propyl ether C₆H₁₄O	.....	.....	.....	.....	5.0	20	70	89.5	-122	LB
n-Hexane C₆H₁₄	.....	.....	.....	(2.2)	(12)	45	151	68.7	-93.35	D
Seven carbon:										
Benzaldehyde C₇H₆O	.....	.....	.....	.....	.....	.....	(1.9)	179	-26	LB
n-Heptane C₇H₁₆	.....	.....	.....	.....	(2.3)	(11)	46	98.4	-90.61	D
Methylcyclohexane C₇H₁₄	.....	.....	.....	.....	(2.7)	12	46	100.9	-126.59	D
Toluene C₇H₈	.....	.....	.....	.....	(1.3)	(6.7)	28	110.6	-94.99	D
Eight carbon:										
Cyclooctatetrene (C₈H₈)	.....	.....	.....	.....	.....	(1.5)	(7.8)	140.56	-4.68	D
n-Octane C₈H₁₈	.....	.....	.....	.....	.....	(2.9)	14	125.67	-56.8	D
Styrene C₈H₈	.....	.....	.....	.....	.....	(1.1)	(6.0)	145.2	-30.63	D
2,2,4-Trimethylpentane C₈H₁₈ (isooctane)	.....	.....	.....	.....	(2.9)	13	49	99.2	-107.38	D

m-Xylene C_8H_{10}	.....	.....	.....	.....	.....	(1.6)	(8.2)	139.1	−47.87	D
o-Xylene C_8H_{10}	.....	.....	.....	.....	.....	(1.3)	(6.6)	144.4	−25.18	D
p-Xylene C_8H_{10}	.....	.....	.....	.....	.....	.....	(8.8)	138.35	13.26	D
Gallium:										
Triethylgallium $(C_2H_5)_3Ga$	.....	.....	.....	.....	.....	1.5	6.8	142.4	−82	LB
Trimethylgallium $(CH_3)_3Ga$	.....	.....	.....	.....	.....	65	218	55.6	−16	LB
Germanium:										
Bromogermane $GeBrH_3$	.....	.....	.....	.....	25	87	272	52	−32	44
Chlorogermane $GeClH_3$	.....	.....	.....	18	73	234	689	(28)	−52	44
Chlorotrifluorogermane GeF_3Cl	.....	.....	.....	101	653	.....	.....	−21	−66	LB
Dichlorogermane $GeCl_2H_2$	.....	.....	.....	2.1	10	39	133	69.5	−68	44
Difluorodichlorogermane GeF_2Cl_2	.....	.....	.....	79	292	.....	.....	−3	−52	LB
Digermane Ge_2H_6	.....	.....	(1.5)	(26)	(89)	(247)	(623)	(31)	*ca.* −109	LB
Dimethylchlorogermane $(CH_3)_2GeHCl$	.....	.....	.....	(2.4)	(9.7)	31	(93)	(89.4)	−76	45
Dimethylgermane $(CH_3)_2GeH_2$	.....	.....	(5.3)	83	(279)	.....	.....	−0.6	−149	45
Fluorogermane $GeFH_3$	.....	.....	.....	.....	.....	358	.....	(15.6)	−22	46
Germane GeH_4	48	182	.....	.....	.....	.....	.....	(−93)	−165	79
Germanium tetrachloride $GeCl_4$	.....	.....	.....	.....	5.6	24	87	85.8	−50	LB
Germanium tetrafluoride GeF_4	.....	.....	$(8.9)_s$	$(337)_s$	.....	.....	.....	$(-35)_s$	.....	47
Methylchlorogermane CH_3GeH_2Cl	.....	.....	.....	(7.7)	26	73	(190)	(70.9)	−101	45
Methyldichlorogermane CH_3GeHCl_2	.....	.....	.....	.....	(2.8)	11	(36)	(113.2)	−62	45
Methylgermane CH_3GeH_3	.....	4.4	48	(470)	.....	.....	.....	(−35.1)	−158	45
Trichlorofluorogermane $GeFCl_3$	.....	.....	.....	(5.7)	(30)	(120)	(435)	37	−50	LB
Trichlorogermane $GeHCl_3$	.....	.....	.....	.....	4.9	24	100	74	−72	LB
Trigermane Ge_3H_8	.....	.....	.....	.....	(2.8)	(11)	(39)	111.1	−106	LB
Halogen										
Bromine Br_2	.....	.....	.....	1.6_s	13_s	66	215	58.8	−7.1	MCA
Bromine pentafluoride BrF_5	.....	.....	.....	8.2	37	130	407	40.9	−60.6	MCA
Bromine trifluoride BrF_3	.....	.....	.....	.....	.....	1.3	7.7	125.8	8.8	MCA
Chlorine Cl_2	.....	2.5_s	44	453	.....	.....	.....	−34.1	−101	MCA

Table of vapor pressures (Continued)

Compound	Temperature, °C							bp(°C)	mp(°C)	Ref.
	−126.59	−111.95	−83.6	−45.21	−22.75	0.00	25.00			
	Vapor pressure, torr									
Chlorine dioxide ClO_2	......	......	......	34	155	485	......	10.9	−59	MCA
Chlorine dioxygen fluoride ClO_2F	......	......	5.0	104	355	......	......	−5.7	−115	MCA
Chlorine fluoride ClF	50	173	......	......	......	......	......	−90	−155.6	MCA
Chlorine pentafluoride ClF_5	......	(1.7)	19	190	533	......	......	−14	−103	78
Chlorine trifluoride ClF_3	......	......	......	33	139	451	......	11.8	−76.3	MCA
Chlorine trioxygen fluoride ClO_3F (perchloryl fluoride)	1.4	7.9	90	......	......	......	......	−46.7	−147.7	MCA
Deuterium chloride DCl	28_s	120	......	......	......	......	......	−84.8	−114.7	MCA
Deuterium fluoride DF	......	......	......	47	146	385	......	18.6	−83.6	MCA
Dichlorine heptoxide (Cl_2O_7)	......	......	......	......	5.5	23	85	84	−91.7	MCA
Dichlorine oxide Cl_2O	......	......	3.4	69	246	696	......	2.1	−120.6	MCA
Dioxygen difluoride F_2O_2	4.5	20	173	......	......	......	......	−56	−163.5	MCA
Fluorine F_2	......	......	......	......	......	......	......	−188	−219.6	MCA
Hydrogen bromide HBr	5.3_s	28_s	299	......	......	......	......	−66.7	−87	MCA
Hydrogen chloride HCl	30_s	124	......	......	......	......	......	−85	−114.2	MCA
Hydrogen fluoride HF	......	......	......	40	131	359	......	19.5	−83.4	MCA
Hydrogen iodide HI	2.0_s	38_s	498	......	......	......	......	−35.6	−50.9	MCA
Iodine heptafluoride IF_7	......	......	1.7_s	44_s	179_s	580_s	......	6	6.0	MCA
Iodine monochloride ICl	......	......	......	......	......	......	27	97.8	13.9	MCA
Iodine pentafluoride IF_5	......	......	......	......	......	4.0_s	26	102	9.4	MCA
Oxygen difluoride F_2O	......	......	......	......	......	......	......	−144.9	−223.8	MCA
Indium: Trimethylindium $(CH_3)_3In$	......	......	......	......	......	......	7.2 at 30°	135.8	88.4	LB

Compound										Ref.
Lead:										
Tetraethylplumbane $(C_2H_5)_4Pb$					1.7	7.6	0.4	183		LB
Tetramethylplumbane $(CH_3)_4Pb$					5.2	28	30	(110)	-27.5	LB
Nitrogen:										
Acetonitrile CH_3CN					6.9	31	(96)	(81.6)	-45.7	LB
Ammonia NH_3			25_s	408				-33.43	-77.7	MCA
Analine C_6H_7N							0.8	184	-6	LB
Benzonitrile C_7H_5N							0.8	191	-38	LB
Carbonyl cyanide $CO(CN)_2$	157	494					124	65.5	< -195	LB
cis-Difluorodiazine N_2F_2								-105.7	51.8	48
Cyanogen bromide $CNBr$					3.7_s	23_s	120_s		-5.2	LB
Cyanogen chloride $CNCl$						456		12.5		LB
Cyanogen fluoride CNF	3.7_s	22_s	332_s					-72.9_s	-74.3	LB
Deuteroammonia ND_3				353				-31.05	-27.9	MCA
Dicyanogen C_2N_2			19_s	163_s	707	235		-21.3	-50	LB
Diethylamine $(C_2H_5)_2NH$			4.7_s	4.0	19			55.7	-92.2	81
Dimethylamine $(CH_3)_2NH$				44	179	562		6.88	-60.5	80
N,N-Dimethylforamide C_3H_7ON							4.1	149	-58	D
1,1-Dimethylhydrazine $(CH_3)_2N_2H_2$				(1.5)	9.3	41	157			LB
1,2-Dimethylhydrazine $(CH_3)_2N_2H_2$					(2.9)	16	70		41	LB
Dinitrogen pentoxide N_2O_5					5.5_s	53_s	413_s		-9.3	MCA
Dinitrogen tetroxide N_2O_4				5.5_s	52_s	263		33	-81	MCA
Ethylamine $C_2H_5NH_2$				26	112	366		21.2		MCA
Ethylenediamine (1,2-diaminoethane) $C_2H_4(NH_2)_2$							12	16.58	8.5	LB
Fluorine nitrate NO_3F	2.1	10	95	731				117.2	-181	MCA
Hydrazine N_2H_4							14	-44	1.5	MCA
Hydrogen azide HN_3				14		183	512	113.5	-80	MCA
Hydrogen cyanide HCN					58		742	36	-13.29	49
Methylamine CH_3NH_2			4.0	92	342	265	25.7	-6.45	-93.5	MCA

Table of vapor pressures (Continued)

Compound	Temperature, °C							bp(°C)	mp(°C)	Ref.
	−126.59	−111.95	−83.6	−45.21	−22.75	0.00	25.00			
	Vapor pressure, torr									
Methyl hydrazine CH₃N₂H₃					(1.9)	(11)	(50)			LB
N-Methylhydroxylamine CH₃NHOH							6.5$_s$	115	42.5	LB
O-Methylhydroxylamine CH₃ONH₂				3.8	21	84	293	48.1		LB
Methyl isothiocyanate CH₃NCS					2$_s$	8$_s$	24$_s$	119	35.5	LB
Methyl thiocyanate CH₃SCN						2.6	12	132.9	−51	LB
Nitric acid HNO₃					2.6	14	63	83	−41.6	MCA
Nitric oxide NO								−151.8	−163.6	MCA
Nitroethane C₂H₅NO₂						4.9$_5$	20	114	−90	LB
Nitrogen trichloride NCl₃					9.0	37	131	(71)	−27	MCA
Nitrogen trifluoride NF₃								−129.06	−206.8	MCA
Nitromethane CH₃NO₂					1.6	8.3	37	101	−29	LB
Nitrosyl chloride NOCl	1.7	13	46$_s$	102	345			−5.4	−59.6	MCA
Nitrosyl fluoride NOF	14$_s$	84$_s$	175					−59.9	−132.5	MCA
Nitrous oxide N₂O								−88.48	−90.8	MCA
Nitryl chloride NO₂Cl			10	201	573			−15.3	−145	MCA
Nitryl fluoride NO₂F	9.1	44	393					−72.4	−166	MCA
Phenyl isocyanate C₆H₅NCO							2.6	165.5		LB
Piperidine C₅H₁₁N					1.7	8.1	32	106		LB
Pyridine C₅H₅N						4.1	20	115	−42	LB
Pyrrole C₄H₅N						(2.0)	(11)	131		LB
Tetrafluorohydrazine N₂F₄	(40)	(109)	(478)					ca. −73		50
Tetramethylhydrazine (CH₃)₄N₂				1.6	8.9	36	126	73		LB
trans-Difluorodiazine N₂F₂	254	733						−111.4	−172	48
Triethylamine (C₂H₅)₃N					5.2	21	78	89.5	−114.7	81

Compound										Ref.
Trifluoroacetonitrile CF₃CN	7.1	33	298	…	…	…	…	-67.7	-144.4	51 LB
Trifluoronitrosomethane CF₃NO	32	115	…	…	…	…	…	-84.5	-117.1	82 LB
Trimethylamine (CH₃)₃N	…	…	…	76	255	680	196	2.9	-72	LB
Trimethylhydrazine (CH₃)₃N₂H	…	…	…	2.5	14	56	…	…	…	
Oxygen:										
Deuterium oxide D₂O	249	…	…	…	…	…	21	101.4	3.82	MCA
Ozone O₃	730	…	…	…	…	4.6	…	-111.3	-251	MCA
Water H₂O	…	…	…	…	…	…	24	100	0.0	API
Phosphorus:										
1,2-Bis(trifluoromethyl) diphosphine P₂H₂(CF₃)₂	…	…	…	(2.0)	11	43	(147)	(69.5)	…	MCA
Bromodifluorophosphine PBrF₂	…	…	12	184	567	…	…	-16	-133.8	MCA
Chlorodifluorophosphine PClF₂	3.6	15	121	1.5	8.0	32	109	-47.3	-164.8	MCA
Dibromofluorophosphine PBr₂F	…	…	…	47	163	448	…	78.3	-115	MCA
Dichlorofluorophosphine PCl₂F	…	2.5	…	19	76	243	710	13.9	-144	MCA
Difluoroiodophosphine PF₂I	…	…	…	…	…	…	…	(26.7)	-93.5	52
Difluorophosphine PF₂H	20	219	…	…	2.1	8.9	32	-64.6	-124	53
Difluorophosphoric acid HPO₂F₂	…	…	…	…	…	…	…	116.4	-96.5	MCA
Dihydrogen phosphorus trifluoride PH₂F₃	…	…	…	55	207	640	…	3.8	-52	54
Dimethylphosphine (CH₃)₂PH	…	…	(2)	34	117	334	…	20.9	-99	LB
Diphosphine P₂H₄	…	…	…	4.1	20	71	224	59	…	MCA
Hydrogen phosphorus tetrafluoride PHF₄	(4.4)	52	…	556	…	…	…	(-39)	…	54
Methylphosphoryl chloride fluoride CH₃POClF	…	…	…	…	…	(1.6)	(9.2)	(130.7)	-22.8	55
Methylphosphoryl difluoride CH₃POF₂	…	13	190	616	(8.5)	36	…	(99.7)	-36.9	55
μ-Oxo-bisdifluorophosphine (PF₂)₂O	…	…	…	…	…	…	…	-18.3	-133.8	56
Phosphine PH₃	54	172	…	…	…	…	…	-87.7	-133.8	MCA
Phosphorus pentabromide PBr₅	…	…	…	…	…	…	4.6₅	111	103.7	MCA

Table of vapor pressures (Continued)

Compound	Temperature, °C							bp(°C)	mp(°C)	Ref.
	−126.59	−111.95	−83.6	−45.21	−22.75	0.00	25.00			
	Vapor pressure, torr									
Phosphorus pentafluoride PF_5	5.3$_s$	48$_s$						−84.5	−93.8	MCA
Phosphorus tribromide PBr_3							2.8	173.2	−40.5	MCA
Phosphorus trichloride PCl_3				9.4		36	120	76.1	−92	MCA
Phosphorus trifluoride PF_3	127	384						101.2	−151.5	MCA
Phosphoryl bromide chloride fluoride $POBrClF$					6.6	28	100	79	......	MCA
Phosphoryl bromide dichloride $POBrCl_2$							9.9	136	13	MCA
Phosphoryl bromide difluoride $POBrF_2$				14	60	199	582	32.1	−84.8	MCA
Phosphoryl chloride difluoride $POClF_2$			3.2	68	240	672		3.1	−96.4	MCA
Phosphoryl dibromide fluoride $POBr_2F$					3.0	12	42	110	−117.2	MCA
Phosphoryl dichloride fluoride $POCl_2F$				4.8	23	84	267	52.9	−80.1	MCA
Phosphoryl trichloride $POCl_3$							37	105.5	1.17	MCA
Phosphoryl trifluoride POF_3			7.6$_s$	465$_s$				−39.5	−39.1	MCA
Tetrafluorodiphosphine P_2F_4			8.1	114	367			(−6.2)	−86.5	57
Tetrakis(trifluoromethyl)cyclotetraphosphine $P_4(CF_3)_4$							2.3$_s$	(135)	66.4	58
Trifluoromethylphosphorus tetrachloride CF_3PCl_4						3.0	13	(104)	−52	58
Trimethylphosphate $(CH_3)_3PO$							(.9)	191	......	LB

304

Compound	1	2	3	4	5	6	7	b.p. (°C)	m.p. (°C)	Ref
Trimethylphosphine $(CH_3)_3P$				13	50	158	459	38.4		LB
Tris(trifluoromethyl)phosphine $(CF_3)_3P$		(3.3)						(17.3)		5
1,2,3-Tris(trifluoromethyl)triphosphine $P_3H_2(CF_3)_3$						3.6	17	ca. 123		59
Radon:										
Radon Rn	(4.1)s	(21)s	(217)s					−62.0	−71	MCA
Selenium:										
Carbon diselenide CSe_2		2.1	24	254		4.4	18	121.8	−45.5	LB
Carbonyl selenide COSe		2.5s	51s	654				−21.9	−124	LB
Hydrogen selenide H_2Se								−42	−65.7	MCA
Methylselenol CH_3SeH		(1.4)	(25)	92				25.4		LB
Selenium hexafluoride SeF_6	1.8s		44s			2.7	16	−46.3s	−34.8	MCA
Selenium tetrafluoride SeF_4							4.2	105	−9.5	MCA
Selenyl fluoride $SeOF_2$						3.2		126	15	MCA
Silicon:										
Allyltrichlorosilane $C_3H_5SiCl_3$							17	(116.8)		60
1,2-Bis(trichlorosilyl)-ethane $(SiCl_3)_2C_2H_4$							(1.3)	(202)		60
Bromodichlorofluorosilane $SiCl_2BrF$			1.7	25	80	216	543	35.3	−112	LB
Bromosilane SiH_3Br			6.2	83	265	707		1.9	−94	61
Bromotrifluorosilane SiF_3Br				658				−41.8	−70.5	LB
i-Butylsilane $(i\text{-}C_4H_9)SiH_3$				(7.1)	(32)	(109)	(322)	(48.6)		62
n-Butylsilane $(n\text{-}C_4H_9)SiH_3$				(4.2)	(22)	(79)	(242)	(56.4)		62
Chlorosilane SiH_3Cl		2.5	36	393	22	73	219	−30.4	−118.1	61
Dibromochlorofluorosilane $SiBr_2ClF$			5.4					59.3	−98	LB
Dibromodifluorosilane SiF_2Br_2		(3.6)	(52)	(167)	(452)			13.7	−67	LB
Dibromosilane SiH_2Br_2			3.0	13	46		142	74.1	−70	LB
Dichlorodifluorosilane SiF_2Cl_2		4.3	44	418				−32.2	−139	LB
Dichlorosilane SiH_2Cl_2		3.8	59	198	548			(8.3)	−122	61

Table of vapor pressures (Continued)

Compound	Temperature, °C							bp(°C)	mp(°C)	Ref.
	−126.59	−111.95	−83.6	−45.21	−22.75	0.00	25.00			
	Vapor pressure, torr									
Diethyldichlorosilane $(C_2H_5)_2SiCl_2$	...	...	...	...	...	(2.6)	(13)	(130.4)	...	60
Diethylsilane $(C_2H_5)_2SiH_2$	...	...	...	(4.8)	23	83	254	...	...	62
Dimethylchlorofluorosilane $(CH_3)_2SiClF$	...	...	...	12	51	171	500	36.2	...	LB
Dimethyldichlorosilane $(CH_3)_2SiCl_2$	...	...	...	(2.3)	11	43	142	70.5	...	60
Dimethyldifluorosilane $(CH_3)_2SiF_2$	...	...	(3.4)	(67)	239	(679)	...	2.1	...	LB
Dimethylsilane $(CH_3)_2SiH_2$	...	1.5	23	253	686	...	...	−20.1	−150	61
Disilane Si_2H_6	...	1.4	18	195	548	...	...	14.5	−132.5	61
Disiloxane $(SiH_3)_2O$	...	...	15	193	563	...	...	−15.2	−143.6	61
Ethyldichlorosilane $C_2H_5SiHCl_2$	...	...	...	2.5	11	38	122	75.5	...	60
Ethylsilane $C_2H_5SiH_3$	...	...	(14)	(182)	(519)	...	...	(−13.7)	−179.7	62
Ethyltrifluorosilane $C_2H_5SiF_3$	...	...	(6.2)	(97)	325	...	...	−5.4	...	LB
Ethnylsilane C_2HSiH_3	...	...	25	263	746	...	...	−22.4	−90.7	63
Hexafluorodisilane Si_2F_6	...	...	...	71s	560s	...	...	−19.1	−18.9	LB
Hexamethyldisiloxane $(CH_3)_6Si_2O$	...	...	...	...	1.7	8.3	(37)	99.2	...	LB
Iododisilane Si_2H_5I	...	...	...	...	3.3	13	45	(102.8)	...	64
Methylchlorofluorosilane $CH_3SiHClF$	...	...	5.8	83	(267)	(722)	...	...	...	LB
Methyldichlorofluorosilane CH_3SiCl_2F	...	...	...	(19)	74	227	643	29.4	...	LB
Methyldichlorosilane CH_3SiHCl_2	...	...	...	11	47	150	429	(40.9)	...	60
Methyldifluorosilane CH_3SiHF_2	...	(3.7)	44	(475)	...	...	...	...	...	LB
Si-Methyldisilazane $HNSiH_3SiH_2CH_3$	...	...	14	...	60	189	544	34.0	...	LB
Methylsilane CH_3SiH_3	4.8	23	195	...	...	...	...	−56.8	−156.5	61
Methyltrichlorosilane CH_3SiCl_3	...	...	...	(2.6)	(14)	(51)	(167)	(66.4)	...	60

Compound										Ref.
Methyltrifluorosilane CH₃SiF₃	304	(1.9)	28	350				−30.2	−185	LB
Silane SiH₄		760						−111.9	−105.6	61
Silazane (SiH₃)₃N					32	109	315	52	5	61
Silicon tetrabromide SiBr₄				7.2	22	77	7.6	153	−68.8	LB
Silicon tetrachloride SiCl₄				(4.8)			237	57.3		60
Silicon tetrafluoride SiF₄	18ₛ	122ₛ			5.9	41		−95ₛ	32	LB
Silylcyanide SiH₃CN			7.7	77	213	506	250	49.6		LB
Silylphosphine SiH₃PH₂			1.1	24	92	(276)		12.1	<−135	LB
Tetramethylsilane (CH₃)₄Si							(738)	26.2	−101.7α / −99.5β	62
Tetrasilane Si₄H₁₀				1.5	1.7	8.1	33	(109)	ca. −90	61
Tribromofluorosilane SiFBr₃				56	7.0	26	86	83.7	−82.5	LB
Tribromosilane SiHBr₃				17	2.2	8.8	32	112	−74	LB
Trichlorofluorosilane SiFCl₃				(4.7)	177	472		12.6	−121	LB
Trichlorosilane SiHCl₃				37	70	218	598	31.8	−126.5	61
Trifluorochlorosilane SiF₃Cl	13	49	358	67				−70	−142	LB
Trimethylchlorosilane (CH₃)₃SiCl			(2.0)		21	76	234	57.6	−90	60
Trimethylfluorosilane (CH₃)₃SiF			4.1		132	388		16.4	−135.9	LB
Trimethylsilane (CH₃)₃SiH				6.5	221	594	285	(6.7)	−117.4	62
Trisilane Si₃H₈					28	96		52.9	−179.1α / −171.6β	61
Vinylsilane C₂H₃SiH₃			(1.5)	(280)	(753)			ca. −22.6		62
Vinyltrichlorosilane C₂H₃SiCl₃			(24)		(4.6)	19	66	(90.6)		60
Sulfur:										
Bis(pentafluorosulfur)peroxide S₂F₁₀O₂				5.0	25	95	305	49.4		65
Bis(trifluoromethyl)disulfide (CF₃)₂S₂				15	58	183	529	(34.6)	−95.4	66
Bis(trifluoromethyl)sulfide (CF₃)₂S		1.3	19	240	737			−22.2		66
Carbon disulfide CS₂				(9.6)	40	127	362	47.3	−111.95	D
Carbon selenosulfide CSSe				(1.2)	(6.5)	(26)	(89)	(85.3)	−85.2	LB

Table of vapor pressures (Continued)

Compound	Temperature, °C							bp(°C)	mp(°C)	Ref.
	−126.59	−111.95	−83.6	−45.21	−22.75	0.00	25.00			
	Vapor pressure, torr									
Carbon subsulfide C_3S_2	2.5						2		0.4	LB
Carbonyl sulfide COS		12	119					−50.2	−138.2	J
Diethyl sulfide $(C_2H_5)_2S$					(3.2)	15	58	92.1	−104.0	D
Dihydrogen disulfide H_2S_2				1.3	7.6	32	111	76	−89.8	MCA
Dimethyl disulfide $(CH_3)_2S_2$					(1.3)	(6.7)	29	109.7	−84.72	D
Dimethyl sulfide $(CH_3)_2S$				(12)	51	167	485	37.3	−98.3	D
Disulfur decafluoride S_2F_{10}					72	231	657	28.9	−92	MCA
Disulfur dichloride S_2Cl_2						1.9	9.5	137.6	−80	MCA
Disulfur difluoride S_2F_2		211						−91	−120.5	MCA
Ethanethiol C_2H_5SH				13	57	184	527	35.0	−121	API
Hydrogen sulfide H_2S	2.9s	16s	200s					−60.3	−85.5	MCA
Methanethiol CH_3SH			(3.0)	61	213	599		5.96	−121	API
Nitrogen sulfur fluoride NSF			3.3	75	263	677		ca. 3	−89	67
Nitrogen sulfur trifluoride NSF_3				307				−27.1	−72.6	67
Pentafluorosulfur hypofluorite SF_5OF			45	482				−35	−86.0	68
Peroxydisulfuryl difluoride $S_2O_6F_2$				(2.2)	12	46	140	67.1	−55.4	68
1-Propanethiol C_3H_5SH				6.1	28	98	154	67.7	−112	API
Pyrosulfuryl difluoride $S_2O_5F_2$					(7.2)	(36)	296	51		MCA
Sulfinyl bromide $SOBr_2$						1.1	6.3	140	−52.2	MCA
Sulfinyl chloride $SOCl_2$				2.0	9.9	37	120	75.6	−104.5	MCA
Sulfinyl chloride fluoride SOClF			2.1	44	159	463		12.3		MCA
Sulfinyl fluoride SOF_2		4.0	64	710				−43.9		MCA
Sulfur dichloride SCl_2					4.5	27	134	59	−78	MCA
Sulfur dinitrogen difluoride SN_2F_2			12	160	471			−11.1	−108	MCA

Compound										Ref.
Sulfur dioxide SO_2	2.1_s			119	416			−10.0	−75.5	MCA
Sulfur hexafluoride SF_6		13_s	4.9_s					−63.9	−50.5	MCA
Sulfur tetrafluoride SF_4		6.1	68	601		5.4_s	74_s	−40	−124	MCA
Sulfur trioxide SO_3(α, stable)			186_s		0.3_s	31_s	233_s		62.2	MCA
Sulfur trioxide SO_3 (β, metastable)					3.4_s	45_s	260		32.5	MCA
Sulfur trioxide SO_3 (γ, metastable)					6.0_s	52	139		16.8	MCA
Sulfur trioxide SO_3 (liq.)					8.5	39	140	44.8		MCA
Sulfuryl bromide fluoride SO_2BrF					8.4	40		428	40	MCA
Sulfuryl chloride SO_2Cl_2				1.7	9.9			69.4	−54.1	MCA
Sulfuryl fluoride SO_2F_2	3.1	14	141					−55.4	−135.8	MCA
Sulfuryl tetrafluoride SOF_4	(1.6)	(8.7)	(98)					−48.5	−99.6	MCA
Tetrahydrothiophene $(CH_2)_4S$						(4.1)	(18)	(121)	−97	LB
Thiophene C_4H_4S					(4.8)	21	80	84.16	−38.2	D
Trifluoromethanethiol CF_3SH		4.8	56	543				−38	(−159)	69
Trifluoromethylsulfuryl fluoride CF_3SO_2F		(1.5)	(20)	240	724			−21.7		LB
Trimethylene sulfide $(CH_2)_3S$					4.6	17	59	95	−73	LB
Tellurium:										
Ditellurium decafluoride Te_2F_{10}					19	69	218	59.0	−33.7	MCA
Hydrogen telluride H_2Te		1.0_s		82	293			−2	−46	MCA
Tellurium hexafluoride TeF_6			25_s	494_s				-38.4_s	−37.8	MCA
Thallium:										
Triethylthallium $(C_2H_5)_3Tl$							2.5	(192)	−63	LB
Trimethylthallium $(CH_3)_3Tl$							ca. 5 at 20°	(147)	38.5	LB
Tin:										
Stannane SnH_4	(4.6)	18	145					−53	−49.7	LB
Tetramethylstannane $(CH_3)_4Sn$					8.7	33	108	78.0		LB
Tin tetrachloride $SnCl_4$				1.7	1.0	5.2	23	124	−33	LB

Table of vapor pressures (Continued)

Compound	Temperature, °C							$bp(°C)$	$mp(°C)$	Ref.
	-126.59	-111.95	-83.6	-45.21	-22.75	0.00	25.00			
	Vapor pressure, torr									
Transition metal:										
Cromyl chloride CrO_2Cl_2	.....	.....	.....	.....	.....	4.3	20	116	-95	LB
Cobalt nitrosyl tricarbonyl $Co(CO)_3NO$	.....	.....	.....	.....	.....	27	95	77.8	(-112)	LB
Hafnium borohydride $Hf(BH_4)_4$	.....	.....	.....	1.4_{81}	.....	2.0_8	15_8	118	28.8	LB
Iridium hexafluoride IrF_6	.....	.....	.....	.....	11_{81}	60_{81}	226_{82}	53.6	43.8	70
Iron dinitrosyl dicarbonyl $Fe(CO)_2(NO)_2$	.....	.....	.....	.....	.....	4.8_8	26	107.6	18.4	LB
Iron pentacarbonyl $Fe(CO)_5$	.....	.....	.....	.....	.....	7.5	30	105	-21	LB
Iron tetracarbonyl dihydride $Fe(CO)_4H_2$	.....	.....	.....	1.8	6.1	17	43	....	-70	LB
Molybdenum hexafluoride MoF_6	.....	.....	.....	5.1_{81}	37_{81}	168_{82}	(557)	34	17.4	70
Nickel tetracarbonyl $Ni(CO)_4$	.....	.....	.....	6.5_8	43_8	135	394	43	-22	71
Osmium hexafluoride OsF_6	.....	.....	.....	1.9_{81}	15_{81}	84_{82}	313_{82}	47.5	33.4	70
Platinum hexafluoride PtF_6	.....	.....	.....	.....	4.0_{81}	26_{81}	112_{82}	69.14	61.3	72
Rhenium hexafluoride ReF_6	.....	.....	.....	4.7_{81}	33_{81}	172_{81}	546	33.8	18.7	70
Rhenium oxypentafluoride $ReOF_5$	.....	.....	.....	.....	3.9_8	22_8	108_8	73	40.8	73
Titanium tetrachloride $TiCl_4$	.....	.....	.....	.....	.....	2.5	(11)	(138)	-23	LB
Tungsten hexafluoride WF_6	.....	.....	.....	14_{81}	87_{81}	373_{82}	....	17.1	2.0	70
Vanadium tetrachloride VCl_4	.....	.....	.....	.....	.....	1.8_8	7.5	152	-26	LB
Zirconium borohydride $Zr(BH_4)_4$	.....	.....	.....	.....	.....	1.0_8	8.2_8	123	28.7	LB
Uranium:										
Uranium hexafluoride UF_6	.....	.....	.....	.....	2.1_8	17_8	110_8	56.5_8	64.1	LB

310

											MCA
Xenon:											
Xenon Xe..............................	191_s	606_s	....	....	....	....	....	....	-108.1	-111.9	74
Xenon difluoride XeF_2.............	....	....	....	....	....	0.6_s	4.5_s	....	....	129	75
Xenon hexafluoride XeF_6............	....	....	....	....	....	2.6_s	29_s	....	....	46	74
Xenon tetrafluoride XeF_4...........	....	....	....	....	....	0.3_s	2.5_s	....	....	117	74
Xenon oxygen tetrafluoride $XeOF_4$....	....	....	....	....	....	7.9	29	....	....	-28	76
							At 23°				
Zinc:											
Diethylzinc $(C_2H_5)_2Zn$............	....	....	....	....	9.1	3.6	16	....	(117.6)	-30	2
Dimethylzinc $(CH_3)_2Zn$.............	....	....	....	....	37	123	371	....	43.8	-29.2	LB
Di-n-propylzinc $(C_3H_7)_2Zn$........	....	....	....	....	....	1.9	8.5	....	(139.4)	-81	2

References for table of vapor pressures:

API refers to the American Petroleum Institute Tables, *D* to Dreisbach's books, *J* to Jordan's book. *MCA* to the Manufacturing Chemists Association Tables, and *LB* to Landolt Bornstein. The complete citations to these sources are given in this appendix under *Sources of Vapor Pressure Data*. In the following references (f) means that the original data was fit by Eq. (V.9).

1. Burg and Schlesinger, *J. Am. Chem. Soc.*, **62**:3425 (1940).
2. Bramford et al., *J. Chem. Soc.*, **1946**:468.
3. Berka et al., *J. Inorg. Nucl. Chem.*, **14**:190 (1960)(f).
4. Sherman and Giauque, *J. Am. Chem. Soc.*, **77**:2154 (1955).
5. Benett et al., *J. Chem. Soc.*, **1953**:1565.
6. Schlesinger, et al., *J. Am. Chem. Soc.*, **62**:3421 (1940).
7. Urry et al., *J. Am. Chem. Soc.*, **76**:5293 (1954).
8. Ceron et al., *J. Am. Chem. Soc.*, **81**:6368 (1959).
9. Burg and Sandhu, *Inorg. Chem.*, **4**:1467 (1965).
10. Urry et al., *J. Am. Chem. Soc.*, **76**:5299 (1954).
11. Stafford and Stone, *J. Am. Chem. Soc.*, **82**:6238 (1960).
12. Stock, "Hydrides of Boron and Silicon," Cornell University Press, Ithaca, N.Y., 1933.
13. Pohland and Harlos, *Z. Anorg. Allgem. Chem.*, **207**:242 (1932).
14. Wirth and Palmer, *J. Phys. Chem.*, **60**:911 (1956).
15. Holtzman et al., "Production of Boranes and Related Research," Table 39D, Academic Press, Inc., New York, 1967.
16. Alton et al., *J. Am. Chem. Soc.*, **81**:3550 (1959).
17. Burg and Randolph, *J. Am. Chem. Soc.*, **71**:3451 (1949).
18. Burg and Schlesinger, *J. Am. Chem. Soc.*, **58**:409 (1936).
19. Schlesinger et al., *J. Am. Chem. Soc.*, **60**:1296 (1938).
20. Burg, *J. Am. Chem. Soc.*, **62**:2228 (1940).
21. *Parsons* et al., *J. Am. Chem. Soc.* **79**:5091 (1957).
22. Egan et al., *Inorg. Chem.*, **3**:1024 (1964).
23. Burg and Kratzer, *Inorg. Chem.*, **1**:725 (1962).
24. Spielman and Burg, *Inorg. Chem.*, **2**:1139 (1963).
25. Wirth and Palmer, *J. Phys. Chem.*, **60**:914 (1956).
26. Burg and Schlesinger, *J. Am. Chem. Soc.*, **55**:4009 (1933).
27. Parry and Bissot, *J. Am. Chem. Soc.*, **78**:1524 (1956).
28. Burg and Spielman, *J. Am. Chem. Soc.*, **81**:3479 (1959).
29. Finch and Schlesinger, *J. Am. Chem. Soc.*, **80**:3573 (1958).
30. Brotherton, et al., *J. Am. Chem. Soc.*, **82**:6245 (1960).
31. Washburn et al., *Adv. in Chem., Ser.*, **23**:129 (1959).
32. Stock and Zelder, *Chem. Ber.*, **54**:531 (1921).
33. Haworth and Hohnstedt, *J. Am. Chem. Soc.*, **82**:3860 (1960).
34. Burg and Randolph, *J. Am. Chem. Soc.*, **73**:953 (1951).
35. Benning and McHarness, *Ind. Eng. Chem.*, **32**:497 (1940).
36. G. F. Furukawa et al., *J. Res. Nat. Bur. Std.*, **51**:69 (1953).
37. Banks et al., *J. Chem. Soc.*, **1948**:2188.
38. Dudley et al., *J. Am. Chem. Soc.*, **70**:3986 (1948) (f).
39. Stull, *Ind. Eng. Chem.*, **39**:517 (1947) (f).
40. Mears et al., *Ind. Eng. Chem.*, **47**:1449 (1955).
41. Booth et al., *J. Am. Chem. Soc.*, **55**:2231 (1933).
42. G. Kubas, unpublished observations, Northwestern University, 1968; Ansul Co. Data Sheet.

43. Kolski and Schaeffer, *J. Phys. Chem.*, **64**:1696 (1960).
44. Dennis and Judy, *J. Am. Chem. Soc.*, **51**:2321 (1929).
45. Amberger and Boeters, *Agnew Chem.*, **73**:114 (1961).
46. Sravastava and Onyszchuk, *Proc. Chem. Soc.*, **1961**:205.
47. Fischer and Weidemann, *Z. Anorg. Allgem. Chem.*, **213**:106 (1933).
48. Colburn et al., *J. Am. Chem. Soc.*, **81**:6397 (1959).
49. Giauque and Ruehrwein, *J. Am. Chem. Soc.*, **61**:2626 (1939).
50. Colburn and Kennedy, *J. Am. Chem. Soc.*, **80**:5004 (1958).
51. Pace and Bobka, *J. Chem. Phys.*, **35**:454 (1961).
52. Rudolph and Parry, *Inorg. Chem.*, **4**:1339 (1965).
53. Rudolph et al., *Inorg. Chem.*, **5**:1464 (1966).
54. Holmes and Storey, *Inorg. Chem.*, **5**:2146 (1966).
55. Zeffert et al., *J. Am. Chem. Soc.*, **82**:3843 (1960).
56. Rudolph et al., *J. Am. Chem. Soc.*, **88**:3729 (1966).
57. Lustig et al., *J. Am. Chem. Soc.*, **88**:3875 (1966).
58. Mahler and Burg, *J. Am. Chem. Soc.* **80**:6161 (1958).
59. Burg and Peterson, *Inorg. Chem.*, **5**:943 (1966).
60. Jenkins, *Ind. Eng. Chem.*, **46**:2367 (1954).
61. Stock, *Z. Elektrochemie*, **32**:341 (1926).
62. Tannenbaum et al., *J. Am. Chem. Soc.*, **75**:3753 (1953).
63. Ebsworth and Frankiss, *J. Chem. Soc.*, **1963**:661.
64. Ward and MacDiarmid, *J. Am. Chem. Soc.*, **82**:2151 (1960).
65. Merrill and Cady, *J. Am. Chem. Soc.*, **83**:298 (1961) (f).
66. Brandt et al., *J. Chem. Soc.*, **1952**:2198.
67. Glemser and Richert, *Z. Anorg. Allgem. Chem.*, **307**:313 (1961) (f).
68. Dudley et al., *J. Am. Chem. Soc.*, **78**:1553 (1956).
69. Dinney and Pace, *J. Chem. Phys.*, **32**:805 (1960).
70. Cady et al., *J. Chem. Soc.*, **1961**:1563.
71. Anderson, *J. Chem. Soc.*, **1930**:1653.
72. Weinstock et al., *J. Am. Chem. Soc.*, **83**:4310 (1961).
73. Cady et al., *J. Chem. Soc.*, **1961**:1568.
74. Schreiner et al., Abstracts of Papers, 153d ACS Meeting, April 1967, Sec. R, no. 127.
75. Weinstock et al., *Inorg. Chem.*, **5**:2189 (1966).
76. Chernick et al., "Noble Gas Compounds," p. 106, University of Chicago Press, Chicago, 1963.
77. Lamb and Roper, *J. Am. Chem. Soc.*, **62**:806 (1940).
78. Pilipovich et al., *Inorg. Chem.*, **6**:1918 (1967).
79. Finholt et al., *J. Am. Chem. Soc.*, **69**:2692 (1947) (f).
80. Aston et al., *J. Am. Chem. Soc.*, **61**:1539 (1939).
81. Pohland and Mehl, *Z. Physik. Chem.*, **A164**:48 (1933).
82. Aston et al., *J. Am. Chem. Soc.*, **66**:1171 (1944).

SOURCES OF VAPOR PRESSURE DATA

Boublik, T. and H. E. Vojitech, 1984, *The Vapor Pressures of Pure Substances, Physical Sciences Data*, Vol. 17, 2nd ed., Elsevier, Amsterdam.

Dreisbach, R. R., 1955, *Physical Properties of Chemical Compounds*, Vols. 1, 2, and 3 (Advances in Chemistry Series, Nos. 15, 22 and 29). American Chemical Society, Washington, D.C. Antoine parameters are given for organic compounds in these books; frequently these have been taken from other compilations.

Gmelin Handbook of Inorganic Chemistry, Springer-Verlag, Berlin. While not specifically devoted to vapor pressures, equations and references to data are available in this multi-volume compendium.

Jordan, T. E., 1954, *Vapor Pressure of Organic Compounds*, Interscience, New York. Vapor pressure equations, tabulations, and graphs for organic and organometallic compounds. This book contains a fair number of errors and inconsistencies.

Stull, D. R., 1947, *Ind. Eng. Chem.*, 39, 517, 1684. Reproduced with minor changes in *Handbook of Chemistry and Physics* and in *Chemical Engineers' Handbook*. Stull presents tabulations of temperatures corresponding to vapor pressures from 1–760 torr for 1,500 compounds. In addition, short tables of data are given for the 1–60-atm region. Even though 40 years old, this is a valuable compilation.

Wilhoit, R., and B. Zwolinski, 1971, *Handbook of Vapor Pressures and Heats of Vaporization of Hydrocarbons and Related Compounds*, College Station, TX: Texas A & M.

PRESSURE AND
FLOW
CONVERSIONS

Conversion factors for various pressure units and flow units are given in Table VI.1. Table VI.2 lists values of the ratio d_t/d_0 (the density of mercury at a temperature t divided by the density of mercury at $0°C$) over the temperature range 0–$99°C$. Capillary depression corrections for mercury in glass tubes are given in Table VI.3. The complete set of corrections for a pressure measurement is made as follows:

$$P(\text{mm}) = h_{\text{obs}}(g/g_0)(d_t/d_0) + C_{\text{u}} - C_{\text{l}}$$

where h_{obs} is the observed difference in height of the two mercury columns, g/g_0 is a correction for the local acceleration due to gravity (where g is the local value of the acceleration due to gravity and $g_0 = 980.665$ cm/s^2), d_t/d_0 is a correction for the density of mercury (see Table VI.2), C_{u} is the capillary depression correction for the upper column of mercury (see Table VI.3), and C_{l} is the capillary depression correction for the lower column (see Table VI.3).

Table VI.1. Conversion Factors for Pressure Units and Flow Units

Pressure units: 1 atm = 760 torr

= 1.01325 bar

= 101,325 Pa (Nm^{-2})

= 1,013,250 dyne/cm^2

= 1033.23 g/cm^2

= 14.6960 lb/$in.^2$

= 760 mm (Hg at 0°C)

Flow units: 1 L/s = 0.06 m^3/min

= 2.11880 ft^3/min

= 15.850342 gal/min

Table VI.2. Ratio of the Density of Mercury at Temperature t Divided by the Density at 0°C [a]

°C	d_t/d_0									
	0	1	2	3	4	5	6	7	8	9
0	1.000000	0.999818	0.999637	0.999456	0.999275	0.999094	0.998912	0.998731	0.998550	0.998369
10	0.998188	0.998007	0.997826	0.997645	0.997465	0.997284	0.997103	0.996923	0.996742	0.996562
20	0.996381	0.996201	0.996021	0.995840	0.995660	0.995480	0.995300	0.995120	0.994939	0.994759
30	0.994580	0.994399	0.994219	0.994040	0.993860	0.993680	0.993501	0.993320	0.993141	0.992961
40	0.992782	0.992602	0.992423	0.992244	0.992064	0.991885	0.991705	0.991526	0.991347	0.991168
50	0.990989	0.990810	0.990630	0.990452	0.990273	0.990093	0.989915	0.989736	0.989557	0.989379
60	0.989200	0.989021	0.988842	0.988664	0.988486	0.988307	0.988128	0.987950	0.987771	0.987593
70	0.987415	0.987237	0.987059	0.986880	0.986702	0.986524	0.986346	0.986168	0.985990	0.985812
80	0.985634	0.985456	0.985278	0.985100	0.984922	0.984744	0.984566	0.984389	0.984211	0.984033
90	0.983856	0.983679	0.983501	0.983323	0.983146	0.982968	0.982791	0.982614	0.982436	0.982259

[a]These density ratios are based on the densities given by P. H. Bigg, *Brit. J. Appl. Phys.*, 15, 1111 (1964). This reference should be consulted for values in the range −20 to 0°C and 100 to 300°C, along with estimated errors.

Table VI.3. Capillary Depression Correction for Mercury in Glass Tubes [a]

Tube diameter (mm)	Meniscus height, mm								
	0.2	0.4	0.6	0.8	1.0	1.2	1.4	1.6	1.8
2	2.52	4.51							
3	1.14	2.13	2.93						
4	0.63	1.22	1.73	2.12	2.47				
5	0.39	0.76	1.10	1.41	1.64	1.84			
6	0.26	0.51	0.75	0.96	1.15	1.30	1.42		
7	0.18	0.37	0.53	0.69	0.82	0.94	1.04	1.13	
8	0.13	0.26	0.38	0.50	0.61	0.70	0.78	0.84	0.90
9	0.10	0.20	0.29	0.38	0.45	0.52	0.59	0.64	0.69
10	0.08	0.12	0.21	0.28	0.34	0.39	0.45	0.49	0.53
11	0.06	0.11	0.16	0.20	0.26	0.31	0.35	0.38	0.41
12	0.04	0.08	0.13	0.16	0.20	0.24	0.27	0.29	0.32
13	0.03	0.07	0.10	0.13	0.16	0.18	0.21	0.23	0.25
14	0.03	0.05	0.07	0.10	0.12	0.14	0.16	0.18	0.19
15	0.02	0.04	0.06	0.08	0.09	0.11	0.12	0.14	0.15
16	0.02	0.03	0.05	0.06	0.07	0.09	0.10	0.11	0.12
17	0.01	0.02	0.04	0.05	0.06	0.07	0.08	0.08	0.09
18	0.01	0.02	0.03	0.04	0.04	0.05	0.06	0.07	0.07
19	0.01	0.01	0.02	0.03	0.03	0.04	0.05	0.05	0.06
20	0.01	0.01	0.02	0.02	0.03	0.03	0.04	0.04	0.04

[a]From G. W. Thomson, *Techniques of Organic Chemistry*, Vol. 1, Pt. 1, 3rd ed., A. Weissenberg, Ed., Interscience, New York, 1959, p. 410.

INDEX